An Introduction to Hu Relationships

An Introduction to Human–Animal Relationships is a comprehensive introduction to the field of human–animal interaction from a psychological perspective across a wide range of themes.

Hollin examines the topic of the relationships between humans and animals as seen in owning a companion animal alongside more indirect relationships such as our approaches to eating meat. The core issues under discussion include the moral and ethical issues raised in using animals for entertainment, in therapy, to keep us safe, and in sports such as horse racing. The justifications for hunting and killing animals as sport and using animals in scientific experimentation are considered. The closing chapter looks to the future and considers how conservation and climate change may influence human–animal relationships.

This key text brings an important perspective to the field of human–animal studies and will be useful to students and scholars in the fields of psychology, sociology, animal welfare, anthrozoology, veterinary science, and zoology.

Clive R. Hollin is Emeritus Professor at the University of Leicester, UK. He has published 25 books, mainly on the topic of criminological psychology. This book is the first in a foray into other areas of psychology.

An Introduction to Human–Animal Relationships

A Psychological Perspective

Clive R. Hollin

LONDON AND NEW YORK

First published 2021
by Routledge
2 Park Square, Milton Park, Abingdon, Oxon OX14 4RN

and by Routledge
605 Third Avenue, New York, NY 10158

Routledge is an imprint of the Taylor & Francis Group, an informa business

© 2021 Clive R. Hollin

The right of Clive R. Hollin to be identified as author of this work has been asserted by him in accordance with sections 77 and 78 of the Copyright, Designs and Patents Act 1988.

All rights reserved. No part of this book may be reprinted or reproduced or utilised in any form or by any electronic, mechanical, or other means, now known or hereafter invented, including photocopying and recording, or in any information storage or retrieval system, without permission in writing from the publishers.

Trademark notice: Product or corporate names may be trademarks or registered trademarks, and are used only for identification and explanation without intent to infringe.

British Library Cataloguing-in-Publication Data
A catalogue record for this book is available from the British Library

Library of Congress Cataloging-in-Publication Data
Names: Hollin, Clive R., author.
Title: An introduction to human-animal relationships : a psychological perspective / Clive R. Hollin.
Description: New York : Routledge, 2021. |
Includes bibliographical references and index.
Identifiers: LCCN 2020049706 (print) |
LCCN 2020049707 (ebook) |
ISBN 9780367277598 (paperback) |
ISBN 9780367277574 (hardback) |
ISBN 9780429297731 (ebook)
Subjects: LCSH: Human-animal relationships. | Human-animal relationships--Psychological aspects.
Classification: LCC QL85 .H654 2021 (print) | LCC QL85 (ebook) |
DDC 590--dc23
LC record available at https://lccn.loc.gov/2020049706
LC ebook record available at https://lccn.loc.gov/2020049707

ISBN: 978-0-367-27757-4 (hbk)
ISBN: 978-0-367-27759-8 (pbk)
ISBN: 978-0-429-29773-1 (ebk)

Typeset in Bembo
by MPS Limited, Dehradun

 Printed in the United Kingdom
by Henry Ling Limited

For

Ebony, Buster, Fly, and Toby

Contents

Acknowledgements viii
Preface ix

Introduction 1

PART I
Animals and psychology 3

1 Animals in psychological research 5

PART II
Mainly of cats and dogs 23

2 Animals as companions 25
3 Pet problems: Aggression 51
4 Pet problems: Anxiety 73
5 Solving pet problems 81

PART III
Humans and animals: Friend or foe? 113

6 Animals amusing and assisting humans 115
7 Animals healers 147
8 Eating, hurting, and killing animals 177
9 Into the Anthropocene 188

Epilogue 191
Index 193

Acknowledgements

I would like to thank Flick for her forbearing every time I said "Do you know what I've read today?" as I prepared to deliver another animal-related snippet. Of which there were many. Thanks to Eva for the heads up on various topics, including the Anthropocene, and to Gregory for several helpful tips. Thanks also to Leo for granting writing-free Tuesday and to TG (see Epilogue) just for being around. I'd like to thank the good folk at Taylor and Francis for taking a punt on this book which is a change from my traditional fare. In particular, the editorial team of Lucy McClune and Akshita Pattiyani gave help and encouragement and for that I'm grateful.

Clive Hollin,
Leicester

Preface

At first glance, this book may appear to be a radical departure from my usual topic of psychology and crime. However, there is rather more of an overlap than might first meet the eye. I became interested in the interplay between psychology and animals when writing about interpersonal violence. Violence is a relatively stable behaviour over time and is directed towards a range of victims. It is therefore not a coincidence that people who act violently towards people are also likely to mistreat animals. I wondered if there anything remarkable about victimisation of animals or whether they are just another casualty of the ubiquitous belligerence characteristic of our species? From here, I started thinking about the wider role that animals have played in psychology and, indeed, in our everyday lives. The list of potential topics in my notes collected under the broad rubric "psychology and animals" grew longer and longer until a template for this book was formed. There's little doubt that I'll have forgotten something along the way, but I hope the contents are at least informative.

A second strand feeding into writing this book came from taking over the teaching of a first-year undergraduate course called *Approaches to Psychology*. In 1972, when I was a first-year undergraduate, this course would have been called *History and Theory*. When preparing my course, I was reminded that animals of various kinds played a pivotal role in the work of several of the great figures in the history of psychology. In contemporary psychology, the use of animals in mainstream psychology has rather gone out of fashion although, of course, the more biologically inclined psychologists conduct some of their work on rodents and other animals.

Finally, a third strand is a highly personal one: as a child I was brought up in a family that did not embrace the idea of keeping pets. (I had an otherwise wonderful childhood, as you ask.) However, my partner's family had dogs (and ponies) and so she has a view of animals as an integral part of everyday life. As soon as it was practically feasible, we had our first dog, Ebony, and we have never been without one since. As parents, we had a highly permissive policy on pets so that both our children were allowed as many pets as they could take responsibility for: over time, their inventory expanded to include stick insects, mice, guinea pigs, rabbits, tropical fish, lizards, a gecko (called Monty who my son acquired when he was about 10 years old and who passed away with much sadness when I was writing this book), and, inevitably, dogs, ponies, and horses. (A line would have been drawn at birds in cages but the need for that discussion never arose.) As adults, they both have pets so childhood experience must count for something!

In terms of nomenclature, throughout the text I use the term "human" to distinguish *Homo sapiens* from other types of animals. This is simply a convenience for ease of reading and should not be taken to imply anthropocentric leanings or some esoteric distinction, such as the presence of a soul, between human and non-human animals on my behalf.

Introduction

The study of animal behaviour, *ethology*, has a long history and some ethological studies have become extremely well known. The Dutch biologist Nikolaas Tinbergen (1907–1988) was concerned with the instinctual way that animals organise their behavioural patterns. In his studies with sticklebacks, he explored the propensity of the male three-spined stickleback (a small, highly territorial, freshwater fish) to attack and defend its territory at the sight of another male. In an elegant series of experiments, Tinbergen demonstrated that the colour red was the instinctual trigger or stimulus for attack: if the underside of a wooden model of a stickleback was redder than that of the real fish, the model would be attacked with greater aggression than a real male (Tinbergen, 1952).

The Austrian ethologist Konrad Zacharias Lorenz (1903–1989) investigated the natural phenomenon of *imprinting*. He showed how, in a critical 13- to 16-hour period after hatching, goslings would imprint on the first moving stimulus they saw. In the natural course of events, this would be the mother duck but Lorenz contrived that he would be the first moving object seen by a clutch of goslings. There are photographs of him being faithfully followed by a gaggle of geese who have imprinted on him. Lorenz shared the 1973 Nobel Prize in Physiology or Medicine with Nikolaas Tinbergen and Karl von Frisch. (Karl von Frisch (1886–1982) was an Austrian ethologist who studied bees and his major work was the translation of the honeybee's waggle dance, which bees use to transmit information to other bees about distant sources of food.)

The writer Paul Theroux (2019) describes the emotional rollercoaster of his relationship with a Muscovy duck, which he called Willy. Theroux states that he was the first moving creature Willy saw and from then on their fate was entwined. Imprinting is indeed a powerful element of nature.

The findings of the early ethologists were seen at the time as potentially being important for understanding human parental behaviour and child development (Vicedo, 2009; Zetterström, 2007). The merging of biology and ethnology with psychology was evident by the 1960s, as illustrated by the work of Wladyslaw Sluckin (1919–1985) on imprinting (Sluckin, 1964), informing an ethologically informed analysis of mother-infant bonding (Herbert, Sluckin, & Sluckin, 1982). This style of translational research, extrapolating from animals to humans, has now faded from fashion. The students who arrive each year at university fresh to the study of psychology will now read a contemporary literature, both theoretical and empirical, which is mainly concerned with people. These new students will focus on people: how we develop from infancy to adulthood, how we interact socially, the intricacies of our personalities, the relationship between brain and behaviour, the mysteries of cognition, and the application of psychology to areas as diverse as anti-social behaviour, the world of work, and mental and physical health.

Nonetheless, at the beginning of psychology as an academic discipline a great deal of pioneering research was conducted with non-human animals. The first section of this book picks out a small number of these classic experimental studies and considers their contribution to psychology.

The second section considers the diversity of relationships we humans have with our fellow creatures. These relationships are not best studied within the narrow confines of the psychological laboratory, rather they are best considered in their natural environment. There are many sides to these relationships. There are settings in which animals are our friends (and we theirs): we live with animals as household pets; we enjoy observing animals in their natural habitat as with, say, bird- or whale-watching; and we train animals to save lives, to assist the physically impaired, and to keep us safe in dangerous environments. Set against these partnerships, we take advantage of animals which are not to their well-being (nor arguably to our dignity as a species). The use of animals as a source of entertainment comes in several forms: there are zoos where animals are caged for our fleeting wonder; circuses where animals perform incongruous tricks to amaze us; television advertising where cute animals entice us to part with our cash; and there are sports, such as horse and greyhound racing, where animals are trained to compete for our excitement and for those who wish to gamble.

If some of the entertainment we derive from animals is relatively benign, then there is much that is not. Our species has little hesitation in the cruel exploitation of animals. The third section raises questions about our choices in which animals we elect to eat and, indeed, whether we wish to eat animals. Human cruelty to animals is evident in the maltreatment of household pets, and the killing of animals in recreational hunting. Animals are also put to use in the laboratory, raising a host of issues surrounding vivisection. The closing pages speculate on what the future may hold and how we humans could try to hold back the impending planetary crisis.

References

Herbert, M., Sluckin, W., & Sluckin, A. (1982). Mother-to-infant 'bonding'. *Journal of Child Psychology and Psychiatry*, *23*, 205–221.

Sluckin, W. (1964). *Imprinting and early learning*. Piscataway, NJ: Transaction Publishers.

Theroux, P. (2019, 20 June). Diary. *London Review of Books*, *41*, no. 12, 40–41.

Tinbergen, N. (1952). The curious behavior of the stickleback. *Scientific American*, 187, 22–27. DOI: 10.1038/scientificamerican1252-22.

Vicedo, M. (2009). The father of ethology and the foster mother of ducks: Konrad Lorenz as expert on motherhood. *Isis*, *100*, 263–291.

Zetterström, R. (2007). The Nobel Prize for the introduction of ethology, or animal behaviour, as a new research field: Possible implications for child development and behaviour: Nobel prizes of importance to Paediatrics. *Acta Paediatrica*, *96*, 1105–1108.

Part I
Animals and psychology

1 Animals in psychological research

The discipline of psychology, at least as taught and practiced in the Western world, has three readily identifiable formative strands. The first is the psychoanalytic tradition of Sigmund Freud (1856–1939) and his followers (Brown, 1961); the second is the establishment by Wilhelm Maximilian Wundt (1832–1920) of the first laboratory for experimental studies in the field of psychology at the University of Leipzig in Germany (Blumenthal, 1985); and the third is the influence of a group of Russian scientists which included the neurologist Vladimir Mikhailovich Bekhterev (1857–1927), the naturalist Vladimir Aleksandrovich Wagner (or *Vagner*, 1849–1934), and the physiologist Ivan Petrovich Pavlov (1849–1936). These early Russian scientists, not constrained by academic boundaries, variously concerned themselves with biology, neurology, physiology, and psychology. The work of the last-named researcher, Ivan Petrovich Pavlov, familiar to generations of psychology students, is where the serious story of animals in psychology begins. However, it is interesting to make a small detour to see what Sigmund Freud had to say about animals.

Freud on animals

In his professional work, Freud had little to say about animals, with the exception of those that appeared in his clients' dreams and fantasies. One of Freud's patients, Sergei Pankejeff (1886–1979), came to Freud with an account of a nightmare experienced on the night before his fourth birthday. In the dream Pankejeff was lying in bed when the window swung open and looking out he saw six or seven white wolves, their gaze fixed upon him, sitting in the tree outside his bedroom. In terror at the wolves' stares, he woke up screaming. Freud's account of the case, known as the *Wolf Man*, became a psychoanalytic classic (Freud, 1918).

In his private life, however, Freud had an evident affection for dogs. In 1925, Freud purchased an Alsatian Shepherd for his daughter's protection on her evening walks through Vienna. The dog was called Wolf (make of that what you will) and became a firm family favourite. Braitman (2014) describes how when Freud was in his mid-70s he acquired two red chows, one of which, called Jofi, became a treasured companion. Jofi was allowed in the consulting room during sessions: Freud held the view that Jofi was a calming influence for patients so that they relaxed and became more candid when she was present.

Freud (1917) gave his views on the human–animal relationship:

In the course of his development towards culture man acquired a dominating position over his fellow-creatures in the animal kingdom. Not content with this supremacy, however, he began to place a gulf between his nature and theirs. He denied the possession of reason to them, and to himself he attributed an immortal soul, and made claims of divine descent which permitted him to annihilate the bonds of community between him and the animal kingdom. (p. 140)

As will be evident as this book unfolds, there are many contemporary examples that lend support to Freud's analysis.

Pavlov's dogs

As recounted in legions of introductory textbooks, the scientific work with a powerful bearing on the emerging discipline of psychology was carried out by the Nobel Prize–winning scientist Ivan Pavlov (Samoilov, 2007). Pavlov was a physiologist and was awarded the 1904 Nobel Prize in Physiology or Medicine "in recognition of his work on the physiology of digestion, through which knowledge on vital aspects of the subject has been transformed and enlarged."

However, it was for reasons other than his physiological research that Pavlov became an important figure in psychology.

Pavlov's research relied upon the measurement of dog's rate of salivation under controlled laboratory conditions. In preparation for eating, a dog salivates as a reflex response to the smell and sight of food. The traditional account is that Pavlov's measurements were disturbed because the dogs were salivating when no food was present but when sounds, such as the clanking of the food pails, associated with food were audible. In a series of experiments in which the presentation of food was repeatedly paired with a stimulus such as a ringing bell Pavlov showed that eventually the bell gained the power to elicit the salivation.

The sequence shown in Figure 1.1 shows the steps in the experiment. The dog's naturally occurring reflex is to salivate when it perceives cues associated with food: there

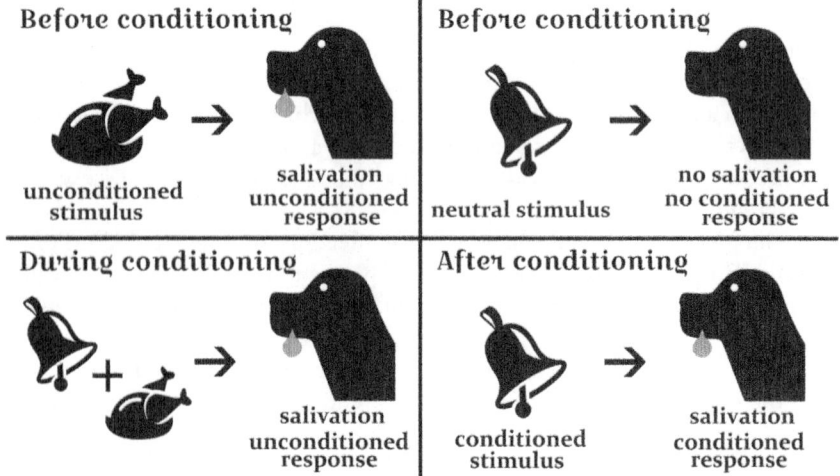

Figure 1.1 Pavlov's experimental design.
Source: Pavlov, I. P. (1897/1902). The work of the digestive glands. London: Griffin.

is no learning involved, thus an unconditioned stimulus (UCS) elicits an unconditioned response (UCR). There is, however, no naturally occurring reason why a dog should salivate at the sound of a bell. In the experiment, the food is repeatedly presented together with the sound of the bell so that the dog learns to associate the food and the sound of the bell. In time, the sound of the bell gains the power to elicit salivation. Thus, the bell is a conditioned stimulus (CS) that elicits salivation as a conditioned response (CR). As dogs do not naturally salivate to the sound of a bell this is sometimes also called a *conditioned reflex*.

The story above is found in the textbooks but, as suggested by Pavlov's biographers, it is not the complete story. While Pavlov used a variety of stimuli, such as a buzzer, harmonium, light, metronome, and whistle, there is some debate about the use of a bell (Thomas, 1997; Todes, 2014). In addition, it appears that some details of terminology may have been lost in translation. The term we favour, *conditionedresponse*, is not what was originally intended: Pavlov used the Russian word *uslovnyi* meaning a *conditionalresponse* (which makes more sense, as the response has become conditional upon the presence of the stimulus).

Pavlov visited the United States in 1923 and 1929 and on the latter visit made presentations at the IXth International Congress of Psychology at Yale University and at the XXXth International Congress of Physiology at Harvard University (Rall, 2016; Ruiz, Sánchez, & De la Casa, 2003). Thus, the early American psychologists, in particular John B. Watson (1878–1958), would have been aware of Pavlov's research and were undoubtably influenced by it (Todd & Morris, 1986).

While Pavlov made no claims to be a psychologist or a behaviourist, he was interested in the use of his experimental methods to understand the mind and consciousness. Pavlov's research was so revolutionary that interest in his work extended beyond academia. The novelist Aldous Huxley was certainly aware of Pavlov's ideas and incorporated them into his seminal work *Brave New World*. In November 1927 the science fiction writer H. G. Wells wrote about Pavlov's life in *The New York Times Magazine*. Thus, Pavlov became a celebrated scientist, recognised by his peers and an international public figure.

In reading the accounts of Pavlov's experiments and their various ramifications it is easy to forget the dogs: however, Tully (2003) provides an excellent canine record. Tully recounts how on a "Pilgrimage to the last working place of the behavioral psychologist Ivan Pavlov in Russia" (p. R117) he discovered a photograph album containing images of some of Pavlov's dogs. These photographs, together with the names of the dogs – Krasavietz, Beck, Milkah, Ikar, Joy, Tungus, Arleekin, Ruslan, Toi, and Murashka – are reproduced in Tully's article.

The use of laboratory dogs was also evident outside psychology as seen, for example, in the development of Beagle Colonies for use in radiation research (Giraud & Hollin, 2016, 2017). Döring, Nick, Bauer, Küchenhoff, and Erhard (2017) found that beagles can be successfully rehomed after life in the laboratory. However, to follow Pavlov's line of work, I have selected a small number of seminal pieces of research which both relied on animals and greatly influenced their own field specifically and psychology generally.

A consideration in the use of laboratory dogs is that they need space and care both of which cost money. Edward Lee Thorndike (1874–1949) had expanded the range of laboratory animals by using cats as well as dogs in his work on the law of effect (Thorndike, 1927). However, as events transpired, it was first pigeons and then rats which proved to be the alternative to dogs. Before looking at the huge part played by the rat in psychological experimentation, we will take a sidestep to look at how animals contributed to Gestalt psychology.

Kohler's chimpanzees

Wolfgang Köhler (1887–1967) was a German psychologist who, along with Max Wertheimer (1880–1943), Fritz Perls (1893–1970), and Kurt Koffka (1886–1941), was a prominent figure in the formation of Gestalt psychology. Gestalt psychology was concerned with how we make sense of our environment. In perceiving the world around we do not focus on every individual element it contains, rather we perceive elements to be part of a greater whole, a gestalt, which can be more than simply the sum of its parts. While no longer a mainstream theory, Gestalt psychology proved to be an important step in the study of human sensation and perception.

Kohler's most well-known work is a series of experimental studies of the problem-solving abilities of chimpanzees, famously with a chimp called Sultan (Kohler, 1925). In one study, a piece of fruit was suspended just out of the chimpanzee's reach and either two sticks or three boxes were placed in close proximity. At first, the chimpanzee tried to jump up to grab the banana but it was too high to reach; after several such failures the chimpanzee attempted to solve the problem. In one study, the problem of getting the banana could be solved by joining the sticks to form a single longer stick to knock down the hanging fruit. In a second study, the chimpanzee solved the problem by stacking the boxes on top of each other and climbing up to reach the fruit (Figure 1.2).

Kohler suggested that that chimpanzees had exhibited a form of learning that he called *insight learning*, the sudden realisation of how to solve a problem. In contrast to trial-and-error learning or learning by observing someone else solve a problem, insight learning is a wholly cognitive process dependent upon being able to visualise the problem and arrive at a solution before making a behavioural response. Of course, once learned, the problem-solving strategy can be repeated when needed in the future. We are all familiar with insight learning: inventions are often the result of insight learning and most people have experienced that *Eureka!* sensation when the solution to a tricky problem "pops into our head."

Skinner's rats (and pigeons)

After dogs, the animal that became widely used in the laboratory was the rat, specifically the albino Norway rat (*Rattus norvegicus*). A much-studied animal (Barnett, 1975), there are several benefits to the use of *Rattus norvegicus* as a laboratory animal: it is clean, easy to breed, and to keep in captivity. From the experimenter's perspective, this breed of rat is easy to tame and to handle and it is a good learner. In all, everything a psychologist could ask for in a rat, although there was some debate about the relative merits of rats born in the laboratory versus rats trapped in the wild (e.g., Boice, 1971; Powell, 1973; Stryjek, 2008).

The first recorded use of laboratory rats in psychological research was by Willard S. Small (1870–1943) at Clark University, Massachusetts. Small's concern was with topics such as the rat's psychic development and its mental processes (Small, 1989, 1900, 1901). However, the most influential experimental studies with rats were carried out at Harvard

Figure 1.2 Kohler's chimpanzees.
Source: Kohler, W. (1924). The mentality of apes. London: Routledge & Kegan Paul.

University by B. F. Skinner (1904–1990). Skinner also conducted research with pigeons but it is his experimental studies of learning in rats for which he is best known.

As a young man, B. F. Skinner was aware of Pavlov's and Thorndike's research, both of which informed his own experiments. Some of Skinner's early wartime work was conducted with pigeons, most notably *Project Pigeon,* which was an attempt to devise a missile guidance system utilising the conditioned pecking behaviour of pigeons (Capshew, 1993; Skinner, 1960). In other research, Skinner investigated topics such as pigeon "superstition" (Skinner, 1948) and self-awareness (Epstein, Lanza, & Skinner, 1981).

Skinner's main body of research was conducted with rats. In order to study the rat's behaviour, he developed a contained environment, an *operant conditioning chamber,* now

10 *Animals and psychology*

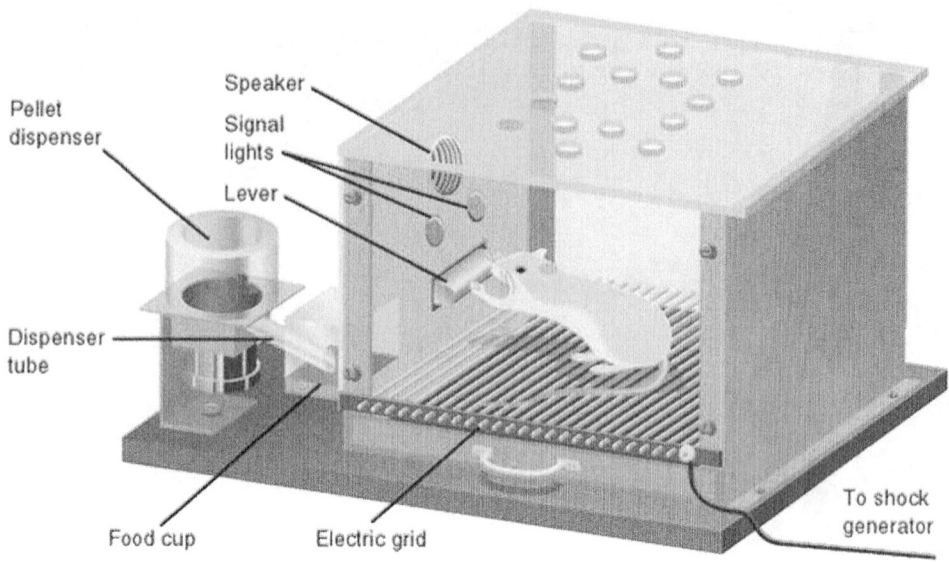

Figure 1.3 Skinner box.
Source: Skinner, B. F. (1938). The behavior of organisms: An experimental analysis. New York: Appleton-Century.

known as a *Skinner box*, which allowed the experimenter to control and manipulate environmental conditions and observe the rat's behaviour (see Figure 1.3).

In a typical experiment a hungry rat is placed in a Skinner box and in the course of exploring its environment it discovers that when it presses a lever a food pellet drops into the food cup. The rat will quickly learn that lever pressing produces the reward of food: in operant terminology, the rat's bar pressing has been *positively* reinforced. There are various experimental manipulations that can be investigated such as the schedule, say fixed versus variable intervals by which rewards are delivered and reinforcement maintained (e.g., Schoenfeld, Cumming, & Hearst, 1956). Skinner defined four types of contingency: (1) *positive reinforcement*, where the frequency of the behaviour is increased or maintained by its rewarding consequences; (2) *negative reinforcement*, where the frequency of the behaviour is increased or maintained by avoiding an aversive consequence; (3) *positive punishment*, where the frequency of the behaviour is decreased by an aversive consequence; (4) *negative punishment*, where the frequency of the behaviour decreases in order to avoid the loss of a reward.

The rat's task can be made more complex by, say, making food available if a light is on but not when the light is off. The rat will learn to lever-press when it is light but not when dark, thereby showing a discrimination between light and dark: the light therefore becomes an Antecedent to the rat's behaviour. The sequence of antecedent : Behaviour : Consequence, correctly called a *three-term contingency*, which emerged from Skinner's experimental analysis of behaviour, provides the framework for the development of applied behaviour analysis. Applied behaviour analysis uses the principles of learning to change behaviours such as delinquency, educational attainment, and mental and physical health (Fisher, Piazza, & Roane, 2013).

As psychology continued to develop through the 1950s and 1960s, the topic of attachment came to prominence. With associations with the work of Lorenz and of Sluckin, the notion of *attachment* refers to a strong emotional bond that can form between two people. John Bowlby (1907–1990) highlighted the importance of the mother–child bond for the infant child's development (e.g., Bowlby, 1953, 1956). However, it was Harlow's research that brought the topic to renewed prominence.

Harlow's monkeys

The American psychologist Harry Harlow (1905–1981), based at the University of Wisconsin–Madison in the state of Wisconsin, was concerned with the nature of the process by which bonding takes place. He conducted a range of studies with newborn rhesus monkeys and their mothers, investigating his thesis that their attachment depends on the mother providing tactile comfort to satisfy the infant's innate need to touch and cling to their mother for emotional comfort.

Harlow used two basic experimental paradigms. In the first, infant monkeys were reared in isolation for varying periods of time, 3, 6, 9, and 12 months, during their first year of life. During their period of isolation, the monkeys behaved abnormally by, for example, grasping their own bodies and rocking impulsively. The isolated monkeys were then placed with other normally socialised monkeys to determine the effects of their failure to form an attachment. When introduced to other monkeys, they showed fear and behaved aggressively, unable to communicate or socialise; they self-harmed, scratching and biting themselves, and were bullied by the other monkeys. The extent of the abnormal behaviour was related to the length of the period of isolation. Those monkeys kept in isolation for 3 months were the least affected, while those held in isolation for 12 months were affected to the point that they failed to recover from the effects of their deprivation (Harlow, Dodsworth, & Harlow, 1965).

In the second experimental procedure, infant monkeys were separated from their mothers immediately after birth and placed for a minimum period of 165 days in cages where they had access to two surrogate mothers. One surrogate mother was made of wire and the other was covered in a soft terry towelling cloth; some monkeys could get milk from the wire mother and some from the cloth mother. The monkeys spent more time with the cloth mother, even if she had no milk, only going to the wire mother when hungry then after feeding returning to the cloth mother. If frightened, the infant monkey sought refuge with the cloth mother (Figure 1.4).

There were later behavioural differences between the monkeys who had grown up with surrogate mothers and normal mothers. The surrogate-reared monkeys were timid, unable to interact with other monkeys, easily bullied, and struggled to mate, while the females became poor mothers. These adverse behaviours were most pronounced in monkeys raised for more than 90 days with a surrogate mother; if placed in a normal environment to allow attachments to other monkeys to form, those with fewer than 90 days exposure were most likely to show recovery of normal functioning.

In all, Harlow concluded that a monkey's normal development relied on some degree of interaction with an object to which they can cling (clinging being a natural response in infant monkeys) during the critical period of the first months of life. The experience of early maternal deprivation caused emotional damage that could be reversed if an attachment was made before the end of the critical period. However, if maternal

12 *Animals and psychology*

Figure 1.4 Harlow's monkeys.
Source: Harlow H. F., Dodsworth R. O., & Harlow M. K. (1965). Total social isolation in monkeys. Proceedings of the National Academy of Sciences of the United States of America.

deprivation continued after the end of the critical period, the emotional damage was permanent (Harlow & Zimmermann, 1959).

Harlow's work demonstrated the nature and permanence of the damage to the infant monkeys that could be caused by maternal and social deprivation. If these findings are generalised to human infants, they reinforce the argument against care homes for babies and favour the view that adoption into a permanent home is the best option.

The ethical issues raised by Harlow's studies are discussed in Chapter 8.

Allen and Beatrix Gardner and Washoe

In 1967 at the University of Nevada, Reno, the psychologists Beatrix (1933–1995) and Allen Gardner, along with primate researcher Roger Fouts, began a project aimed at teaching American Sign Language to a chimpanzee. In 1970, the project moved to the Institute of Primate Studies in Norman, Oklahoma. The previous attempts to teach language to chimpanzees had not been successful (e.g., Hayes & Hayes, 1952), which may have been due to trying to teach the spoken word. The Gardners reasoned that verbal communication may be too difficult for a chimpanzee and they elected to use sign language.

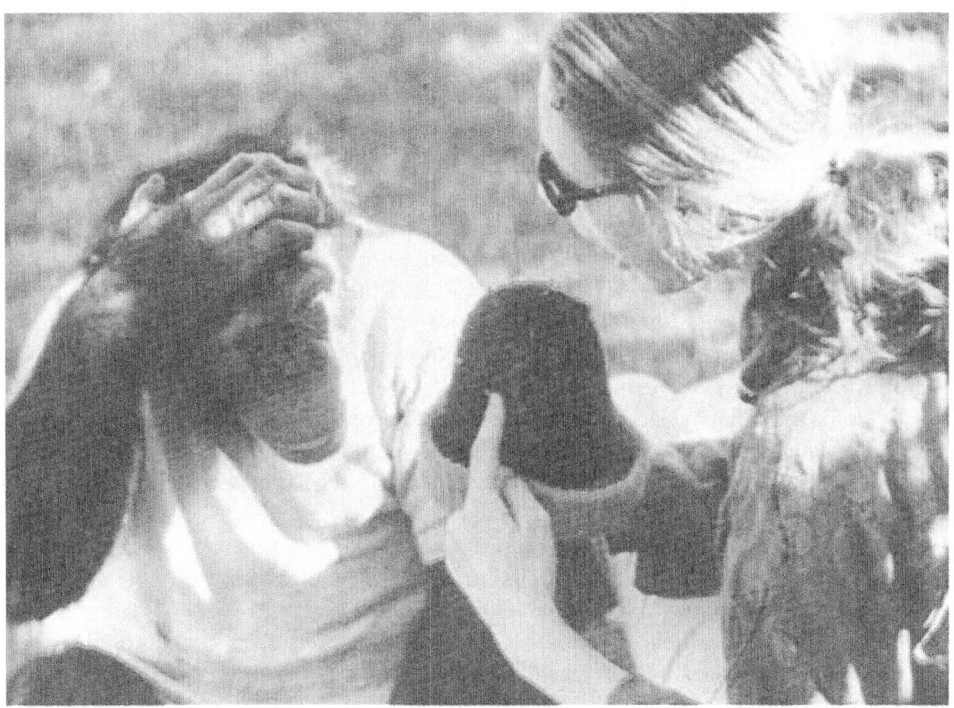

Figure 1.5 Washoe in conversation.
Source: This Photo by Unknown Author is licensed under CC BY-NC-SA. Gardner, RA, Gardner, BT (1969). Teaching sign language to a chimpanzee. Science 165, 664–672.

Their project was carried out the with a female chimpanzee named Washoe, after Washoe County, Nevada. To meet her need for companionship, Washoe (1965–2007) was brought up in an environment as similar as possible to that of a human child. Washoe had her own 8 × 24-foot trailer with spaces for cooking, living, and sleeping. She sat with the family at the dinner table and had access to clothes, toys, books, and so on. Like any human child, she had a regular routine of responsibilities, play, and rides in the family car (Figure 1.5).

The rule within the project was that anyone in the presence of Washoe had to communicate using sign language, rather than speech, in order to create a consistent, less confusing environment for Washoe. Over the duration of the research, Washoe learned over 300 words that she could reliably sign and use appropriately (Gardner & Gardner, 1969; Gardner, Gardner, & Van Cantfort, 1989). The ability to communicate allowed the researchers an appreciation of Washoe's deeper level of understanding about herself and her environment. Thus, for example, Washoe was shown herself in a mirror, and asked what she was saw: she signed "Me, Washoe." It was also observed that Washoe showed empathy for the students working on the project by signing more slowly for newcomers. The full and fascinating story of Washoe's life is recorded in detail on the Friends of Washoe website (friendsofwashoe.org). However, the research into ape language faded such that it became, as summed up by Kulick (2017), a "Promising field that tanked" (p. 359).

Seligman's dogs

What do we do when we are faced with an aversive situation? The natural, instrumental response is to try to escape from or avoid the unpleasant situation, but what happens if there is no escape? In a series of experiments, the psychologist Martin Seligman from the University of Pennsylvania studies the behaviour of dogs unable to escape from or avoid an unpleasant stimulus. The basic experimental setup, as shown in Figure 1.6, is that a dog is placed in a container, called a *shuttle box*, which is divided in two by a fixed barrier. When the dog first receives an electric shock through the floor of the box it reacts by barking, running, urinating, and showing other signs of fear until it jumps the barrier and escapes the shock. This is the standard procedure in an escape-avoidance experimental paradigm. When the procedure is repeated for the next trial, the same dog will cross the barrier more quickly than on the preceding trail and so on for subsequent trials until optimum performance is reached.

The dog's natural avoidance of pain can be interrupted by restraining the dog and exposing it to inescapable electric shocks before beginning the avoidance learning procedure (e.g., Seligman & Maier, 1967). Seligman (1972) describes what happens in this situation:

> Such a dog's first reactions to shock in the shuttle box are much the same as those of a naive dog. However, in dramatic contrast to a naive dog, a typical dog which has experienced uncontrollable shocks before avoidance training soon stops running and howling and sits or lies, quietly whining, until shock terminates. The dog does not cross the barrier and escape from shock. Rather, it seems to give up and passively

Figure 1.6 Seligman's dogs.
Source: Seligman MEP, Maier SF. Failure to escape traumatic shock. Journal of Experimental Psychology. 1967;74:1–9.

accepts the shock. On succeeding trials, the dog continues to fail to make escape movements and takes as much shock as the experimenter chooses to give (p. 407).

The interference effect of prior unescapable shock on normal responding appears to dissipate over time so that after a few days the dog returns to normal functioning.

An explanation for the dog's actions is that it learns that its behaviour and the anticipated consequences (escaping the shock) does not happen when the shock is unavoidable. This experience acts to destabilise the dog's natural instrumental behaviour: the term *learned helplessness* was coined for this type of behaviour (Maier, Seligman, & Solomon, 1969). Later work showed that learned helplessness was not peculiar to dogs but was also found in cats, fish, mice, and rats. This phenomenon has also been observed in both young (Nolen-Hoeksema, Seligman, & Girgus, 1986) and adult humans (Hiroto & Seligman, 1975).

The applied dimension to learned helplessness lay in the parallels between what is observed in animals and in the cognitive, emotional, and motivational features of depression in humans. In particular, the view expressed by some people with depression that they are unable to control significant aspects of their life resonates with the notion of learned helplessness. This observation helped to stimulate a line of clinical research (Abramson, Seligman, & Teasdale, 1978; Maier & Seligman, 2016; Miller & Seligman, 1975).

As psychology matured, the use of animals in psychological experiments fell out of fashion and the emphasis shifted to human cognition. This change is best exemplified with learning theory, the bastion of animal experimentation and the advent of social learning theory with its focus on internal processes such as cognition and emotion (Bandura, 1977). Indeed, the change in direction taken by psychological research was heralded as a cognitive revolution (Baars, 1986). Although there was a renewed interest in *comparative psychology*, a hybrid of psychology and ethology (Greenberg & Haraway, 2002).

Comparative psychology

The notion of comparative psychology has been with us for some time (Morgan, 1902). Dewsbury (2003) notes that contemporary comparative psychology is the study of the functioning of non-human animals which has its roots in several traditions, namely: (i) *European Ethology* exemplified by the work of Lorenz and Tinbergen as discussed above; (ii) *Sociobiology*, which seeks to understand social behaviour in evolutionary terms (Wilson, 1975) extending to *Behavioural Ecology*, the study of influence of ecological forces on the evolution of animal behaviour (Davies, Krebs, & West, 2012); (iii) *Evolutionary Psychology*, the product of merging psychology with evolutionary biology (Dunbar & Barrett, 2007).

The lines of investigation followed by comparative psychology, sometimes referred to as *animal psychology*, may include comparisons across species but that is not its sole purpose. The wider focus of comparative psychology includes heredity and the relative influence on behaviour of genes and environment, mating behaviours, and parenting, and social behaviours such as play, aggression, and communication. At an individual level, the concern may be with topics such as instincts, learning, and eating.

Animal cognition

Following the "cognitive revolution" in mainstream psychology, the study of animal cognition became a focus within comparative psychology. It is clear that while the human brain followed its particular evolutionary path in terms of size and architecture (Holloway, 2015), it has features in common with other animals, particularly the great apes. As exemplified by Washoe, there is some overlap in the cognitive abilities of humans and primates; for example, both species have causal cognition (Penn & Povinelli, 2007), reasoning skills (Vonk & Subiaul, 2009), and the ability to communicate (Moore, 2016). Suddendorf and Whiten (2001) suggest that the level of cognition reached by great apes is similar to that of a 2-year-old human. However, as Vonk and Aradhye (2015) explain, in comparing humans and primates, the unresolved question is whether the similarities and differences in cognitive functioning are simply a matter of degree or whether there is a fundamental gap.

The level of sophistication of cognitive functioning raises the issues of *metacognition* and *consciousness*: the ability to be aware of one's own thoughts and emotions. Is the ability to be aware of one's own cognitions a uniquely human attribute? While it is doubtful that the same degree of metacognition is present in humans and primates, it remains a possibility that some facets of metacognitive ability do cross species boundaries (Smith, Coutinho, Boomer, & Beran, 2012).

However, moving away from primates, there are other species which have attracted attention because of their ostensible cognitive ability. There are several species of birds that display intelligent behaviour (Emery, 2006). In particular, the corvids have a range of cognitive skills (e.g., Bugnyar & Kotrschal, 2002) and it may be a mistake to underestimate the humble chicken (Marino, 2017). A range of species, including primates, several types of birds, and otters use tools (Emery & Clayton, 2009); while away from mammals and birds, lizards display cognitive abilities (Matsubara, Deeming, & Wilkinson, 2017) and fish appear to process social information (Webster & Laland, 2017).

Animal models

The notion of comparative physiology medicine is based on the observation that, to a greater or lesser degree, humans share physiological and behavioural characteristics with other species of animals. It follows that we humans can learn about ourselves by studying animals, particularly those with similar biological functioning. In some instances, the research is harmful to the animals in the study; this point is discussed further in Chapter 8. In tracing the history of using animals to model human functioning, Ericsson, Crim, and Franklin (2013) note that there is nothing new about the notion of animal models; for example, ancient Greeks used dogs to search for the location of intelligence while in the 17th century William Harvey carried out anatomical studies with live animals, including fish and birds, to inform his mapping of human blood circulation.

Geyer and Markou (1995) make the point that in practice the phrase "animal models" has a diversity of meaning. Thus, with reference to animal models of psychiatric disorder, they state that:

> At one extreme one can attempt to develop an animal model that mimics a psychiatric syndrome in its entirety .... At the other extreme, one more limited purpose for an animal model is to provide a way to systematically study the effects of potential therapeutic treatments. (p. 787)

The previously discussed learned helplessness model of depression provides an example of an animal model of a human psychiatric condition. From the original premise, the research progressed and the animal models of depression become more complex, encompassing a wider range of factors (e.g., Czéh, Fuchs, Wiborg, & Simon, 2016). There are animal-based models of other psychiatric conditions that may be so exact as to focus on a particular aspect of a complex disorder such as schizophrenia (Ayhan, McFarland, & Pletnikov, 2016). Alongside physiological and psychiatric conditions, there are also animal models, often incorporating a great deal of biological research, of human behaviours such as conduct disorder (Macrì, Zoratto, Chiarotti, & Laviola, 2018), alcoholism (Higley & Linnoila, 1997), and aggression (de Boer, 2018). The purpose of animal models of aggression is described by de Boer (2018): "Circuit-level knowledge of the neuromolecular underpinnings of escalated aggression has great potential to guide the rational development of effective therapeutic interventions for pathological social and aggressive behavior in humans" (p. 86).

In summary, it is important to keep in mind that a model is just that: a representation of how biological, psychological, and behavioural systems may function and not an exact copy.

Anthropomorphism

One of the hallmarks of familiarity with animals, particularly among pet owners, is a psychological tendency to attribute human properties to an animal in order to explain its behaviour. This is not to say that animals do not have identifiable personalities – there is a body of research concerned with animal personality (e.g., Gartner, 2015) – rather that the cognitive abilities of animals can be over-estimated. Shettleworth (2010) takes the phenomenon of insight learning to illustrate this point. As discussed above, Kohler studied the problem-solving abilities of chimpanzees and explained what he observed in terms of insight learning. However, while this explanation fits the observations, it is not necessarily correct. Shettleworth cites studies showing that pigeons can behave in the same way as chimpanzees in solving the problem of obtaining an out-of-reach reward (e.g., Epstein et al., 1981). An alternative explanation to insight is that it is the animal's accumulated experience with the elements of the problem, rather than a sudden insight, which allow it to behave effectively in the novel, problematic situation. Shettleworth makes the comment that: "Although the extent of human–animal cognitive similarity is undoubtedly a key issue for comparative psychology, it sometimes seems the agenda is to support anthropomorphic interpretations rather than to pit them experimentally against well-defined alternatives" (p. 478).

Shettleworth acknowledges that it is easy to accept a "clever animal" explanation rather than the altogether less wonderful "killjoy" account that seems to deny any significant continuity between human and animal functioning. In support of this position Shettleworth (2012) makes the case that the two types of explanation for animal functioning have their roots in different scientific traditions. A Darwinian perspective has as its core the continuity of evolution so that that there are similarities, including likenesses in mental events, across species. The alternative experimental approach, as seen for example with Tinbergen and with Skinner, is altogether more prosaic in seeking to establish whether such similarities exist.

Serpell (2002) suggests that anthropomorphism can be looked at from an evolutionary perspective. As the social environment of humans and some animals became increasingly shared so it suited both species to maintain that closeness. While the function of the relationship may have changed over time, producing a context where pets are commonplace, the animals provide much valued non-human social support. In return for providing social support the animals receive food, warmth, and an accepting social environment. From this point of view, anthropomorphism is a good thing all round. However, the tipping point from a supportive to a destructive relationship has been passed for some people who keep animals as pets. First, as discussed in the following chapter, the animal's environment becomes one where animals are treated like humans, sometimes to the extreme in terms of wearing human clothes and eating human food. This change may become counterproductive so that an animal used to a great deal of attention becomes distressed, noisy, and destructive when left alone. Second, the animal's appearance may be changed either through surgery, as in docking tails and ears, or through selective breeding. A programme of in-breeding may produce a desired set a of characteristics, such as the dachshund's elongated body, but there may be a physiological price to pay as seen with the bulldog's chronic respiratory problems. In this light, Serpell makes a telling point: "If bulldogs were the products of genetic engineering by agri-pharmaceutical corporations, there would be protest demonstrations throughout the Serpell (2002) Western world, and rightly so. But because they have been generated by anthropomorphic selection, their handicaps not only are overlooked but even, in some quarters, applauded" (p. 447).

The use of animals in psychological research has led to some notable findings, such as the principles of reinforcement and the notion of learned helplessness. However, some of this research also raises ethical and questions about the treatment of the animals in the studies. In the following chapters, the psychological findings will be referred to as the need arises, the moral and ethical issues are looked at in detail later. The focus now shifts from the rarefied world of the psychological laboratory to the seemingly more mundane world of those animals we choose to keep with us as pets.

References

Abramson, L. Y., Seligman, M. E. P., & Teasdale, J. D. (1978). Learned helplessness in humans: Critique and reformulation. *Journal of Abnormal Psychology*, 87, 49–74.
Ayhan, Y., McFarland, R., & Pletnikov, M. V. (2016). Animal models of gene–environment interaction in schizophrenia: A dimensional perspective. *Progress in Neurobiology*, 136, 1–27.
Baars, B. J. (1986). *The cognitive revolution in psychology*. New York: The Guilford Press.
Bandura, A. (1977). *Social learning theory*. Englewood Cliffs, NJ: Prentice-Hall.
Barnett, S. A. (1975). *The rat*. Chicago, IL: University of Chicago Press.
Blumenthal, A. L. (1985). Wilhelm Wundt: Psychology as the propaedeutic science. In C. E. Buxton (Ed.), *Points of view in the modern history of psychology* (pp. 19–50). Orlando, FL: Academic Press.
Boice, R. (1971). Laboratorizing the wild rat (*Rattus norvegiens*). *Behavior Research Methods & Instrumentation*, 3, 177–182.
Bowlby, J. (1953). *Child care and the growth of love*. Harmondsworth, Middlesex: Penguin Books.
Bowlby, J. (1956). Mother-child separation. *Mental Health and Infant Development*, 1, 117–122.
Braitman, L. (2014). Dog complex: Analyzing Freud's relationship with his pets. Fast Company: https://www.fastcompany.com/3037493/dog-complex-analyzing-freuds-relationship-with-his-pets. (Accessed 10 March 2019).
Brown, J. A. C. (1961). *Freud and the post-Freudians*. Harmondsworth, Middlesex: Penguin.
Bugnyar, T., & Kotrschal, K. (2002). Observational learning and the raiding of food caches in ravens, *Corvus corax*: Is it 'tactical' deception? *Animal Behaviour*, 64, 185–195.

Capshew, J. H. (1993). Engineering behavior: Project Pigeon, World War II, and the conditioning of B. F. Skinner. *Technology and Culture, 34*, 835–857.

Czéh, B., Fuchs, E., Wiborg, O., & Simon, M. (2016). Animal models of major depression and their clinical implications. *Progress in Neuro-Psychopharmacology and Biological Psychiatry, 64*, 293–310.

Davies, N. B., Krebs, J. R., & West, S. A. (2012). An introduction to *behavioural ecology* (4th ed.). Chichester, West-Sussex: Wiley-Blackwell.

de Boer, S. F. (2018). Animal models of excessive aggression: Implications for human aggression and violence. *Current Opinion in Psychology, 19*, 81–87.

Dewsbury, D. A. (2003). Comparative psychology. In D. K. Freedheim & I. B. Weiner (Eds.), *Handbook of psychology: Vol. 1: History of psychology* (pp. 67–84). Hoboken, NJ: John Wiley & Sons.

Döring, D., Nick, O., Bauer, A., Köchenhoff, H., & Erhard, M. H. (2017). How do rehomed laboratory beagles behave in everyday situations? Results from an observational test and a survey of new owners. *PLoS ONE, 12*, e0181303.

Dunbar, R. I. M., & Barrett, L. (Eds.). (2007). *The Oxford handbook of evolutionary psychology*. Oxford: Oxford University Press.

Emery, N. J. (2006). Cognitive ornithology: The evolution of avian intelligence. *Philosophical Transactions of the Royal Society B, 361*, 23–43.

Emery, N. J., & Clayton, N. S. (2009). Tool use and physical cognition in birds and mammals. *Current Opinion in Neurobiology, 19*, 27–33.

Epstein, R., Lanza, R. P., & Skinner, B. F. (1981). "Self-awareness" in the pigeon. *Science, 212*(4495), 695–696.

Ericsson, A. C., Crim, M. J., & Franklin, C. L. (2013). A brief history of animal modeling. *Missouri Medicine, 110*, 201–205.

Fisher, W. W., Piazza, C. C., & Roane, H. S. (Eds.). (2013). *Handbook of applied behavior analysis*. New York: Guildford Press.

Freud, S. (1917/1948). A difficulty in the path of psychoanalysis. In *The Standard Edition of the Complete Works of Sigmund Freud*, vol. XVII (pp. 135–144). London: Hogarth Press. (Trans James Strachey.)

Freud, S. (1918/2002). *The 'Wolfman' and other cases*. Harmondsworth, Middlesex: Penguin Modern Classics. (Trans Louise Adey Huish.)

Gardner, R. A., & Gardner, B. T. (1969). Teaching sign language to a chimpanzee. *Science, 165*, 664–672.

Gardner, R. A., Gardner, B. T., & Van Cantfort, T. E. (Eds.). (1989). *Teaching sign language to chimpanzees*. Albany, NY: SUNY Press.

Gartner, M. C. (2015). Pet personality: A review. *Personality and Individual Differences, 75*, 102–113.

Geyer, M. A., & Markou, A. (1995). Animal models of psychiatric disorders. In F. E. Bloom and D. J. Kupfer (Eds.), *Psychopharmacology: The fourth generation of progress* (pp. 787–798). New York: Raven Press.

Giraud, E., & Hollin, G. J. S. (2016). Care, laboratory beagles and affective utopia. *Theory, Culture & Society, 33*, 27–49.

Giraud, E., & Hollin, G. J. S. (2017). Laboratory beagles and affective co-productions of knowledge. In M. Bastian, O. Jones, N. Moore, & E. Roe (Eds.), *More-than-human participatory research* (pp. 163–177). New York: Routledge.

Greenberg, G., & Haraway, M. M. (2002). *Principles of comparative psychology*. Boston, MA: Allyn and Bacon.

Harlow, H. F., & Zimmermann, R. R. (1959). Affectional responses in the infant monkey. *Science, 130*, 421–432.

Harlow, H. F., Dodsworth, R. O., & Harlow, M. K. (1965). Total social isolation in monkeys. *Proceedings of the National Academy of Sciences, 54*, 90–97.

Hayes, K. J., & Hayes, C. (1952). Imitation in a home-raised chimpanzee. *Journal of Comparative and Physiological Psychology, 45*, 450–459.

Higley, J. D., & Linnoila, M. (1997). A nonhuman primate model of excessive alcohol intake. Personality and neurobiological parallels of type I- and type II-like alcoholism. *Recent Developments in Alcoholism, 13*, 191–219.

Hiroto, D. S., & Seligman, M. E. P. (1975). Generality of learned helplessness in man. *Journal of Personality and Social Psychology, 31*, 311–327.

Holloway, R. L. (2015). Brain evolution. In M. P. Muehlenbein (Ed.), *Basics in human evolution* (pp. 235–250). London: Academic Press.

Kohler, W. (1925). *The mentality of apes.* (Translated from the 2nd. revised edition by Ella Winter.) London: Kegan Paul.

Kulick, D. (2017). Human–animal communication. *Annual Review of Anthropology, 46*, 357–378.

Macrì, S., Zoratto, F., Chiarotti, F., & Laviola, G. (2018). Can laboratory animals violate behavioural norms? Towards a preclinical model of conduct disorder. *Neuroscience & Biobehavioral Reviews, 91*, 102–111.

Maier, S. F., & Seligman, M. E. P. (2016). Learned helplessness at fifty: Insights from neuroscience. *Psychological Review, 123*, 349–367.

Maier, S. F., Seligman, M. E. P., & Solomon, R. L. (1969). Pavlovian fear conditioning and learned helplessness. In B. A. Campbell & R. M. Church (Eds.), *Punishment* (pp. 299–342). New York, NY: Appleton-Century-Crofts.

Marino, L. (2017). Thinking chickens: A review of cognition, emotion, and behavior in the domestic chicken. *Animal Cognition, 20*, 127–147.

Matsubara, S., Deeming, D. C., & Wilkinson, A. (2017). Cold-blooded cognition: New directions in reptile cognition. *Current Opinion in Behavioral Sciences, 16*, 126–130.

Miller, W. R., & Seligman, M. E. P. (1975). Depression and learned helplessness in man. *Journal of Abnormal Psychology, 84*, 228–238.

Moore, R. (2016). Meaning and ostension in great ape gestural communication. *Animal Cognition, 19*, 223–231.

Morgan, C. L. (1902). *An introduction to comparative psychology.* New York: Scribner's. (Original work published 1894 by Walter Scott Publishing, New York.)

Nolen-Hoeksema, S., Seligman, M. E. P., & Girgus, J. S. (1986). Learned helplessness in children: A longitudinal study of depression, achievement, and explanatory style. *Journal of Personality and Social Psychology, 51*, 435–442.

Penn, D., & Povinelli, D. (2007). Causal cognition in human and nonhuman animals: A comparative, critical review. *Annual Review of Psychology, 58*, 97–118.

Powell, R. E. (1973). Laboratory study of wild rats. *Bulletin of the Psychonomic Society, 1*, 119–120.

Rall, J. A. (2016). The XIIIth International Physiological Congress in Boston in 1929: American physiology comes of age. *Advances in Physiology Education, 40*, 5–16.

Ruiz, G., Sánchez, N., & De la Casa, L. G. (2003). Pavlov in America: A heterodox approach to the study of his influence. *Spanish Journal of Psychology, 6*, 99–111.

Samoilov, V. O. (2007). Ivan Petrovich Pavlov (1849–1936). *Journal of the History of the Neurosciences, 16*, 74–89.

Schoenfeld, W. N., Cumming, W. W., & Hearst, E. (1956). On the classification of reinforcement schedules. *Proceedings of the National Academy of Sciences of the United States of America, 42*, 563.

Seligman, M. E. P. (1972). Learned helplessness. *Annual Review of Medicine, 23*, 407–412.

Seligman, M. E. P., & Maier, S. F. (1967). Failure to escape traumatic shock. *Journal of Experimental Psychology, 74*, 1–9.

Serpell, J. A. (2002). Anthropomorphism and anthropomorphic selection — beyond the "cute response". *Society & Animals, 10*, 437–454.

Shettleworth, S. J. (2010). Clever animals and killjoy explanations in comparative psychology. *Trends in Cognitive Sciences, 14*, 477–481.

Shettleworth, S. J. (2012). Darwin, Tinbergen, and the evolution of comparative cognition. *The Oxford handbook of comparative evolutionary psychology* (pp. 529–546). Oxford: Oxford University Press.

Skinner, B. F. (1948). 'Superstition' in the pigeon. *Journal of Experimental Psychology, 38*, 168–172.

Skinner, B. F. (1960). Pigeons in a pelican. *American Psychologist, 15*, 28–37.
Small, W. S. (1900). An experimental study of the mental processes of the rat. *American Journal of Psychology, 11*, 133–165.
Small, W. S. (1901). Experimental study of the mental processes of the rat. II. *American Journal of Psychology, 12*, 206–239.
Small, W. S. (1989). Psychic development of the young white rat. *American Journal of Psychology, 11*, 80–100.
Smith, J. D., Coutinho, M. V. C., Boomer, J., Beran, M. J. (2012). Metacognition across species. In J. Vonk & T. K. Shackelford (Eds.), *The Oxford handbook of comparative evolutionary psychology* (pp. 274–291). New York: Oxford University Press.
Stryjek, R. (2008). Devices for handling small mammals in laboratory conditions. *Acta Neurobiologiae Experimentalis, 68*, 407–413.
Suddendorf, T., & Whiten, A. (2001). Mental evolution and development: Evidence for secondary representation in children, great apes and other animals. *Psychological Bulletin, 127*, 629–650.
Thomas, R. K. (1997). Correcting some Pavloviana regarding "Pavlov's bell" and Pavlov's "mugging". *American Journal of Psychology, 110*, 115–11125.
Thorndike, E. L. (1927). The law of effect. *American Journal of Psychology, 39*, 212–222.
Todd, J. T., & Morris, E. K. (1986). The early research of John B. Watson: Before the behavioral revolution. *The Behavior Analyst, 9*, 71–88.
Todes, D. P. (2014). *Ivan Pavlov: A Russian life in science.* Oxford: Oxford University Press.
Tully, T. (2003). Pavlov's dogs. *Current Biology, 13*, R117–R119.
Vonk, J., & Aradhye, C. (2015). Evolution of cognition. In M. P. Muehlenbein (Ed.), *Basics in human evolution* (pp. 479–491). London: Academic Press.
Vonk, J., & Subiaul, F. (2009). Do chimpanzees know what others can and cannot do? Reasoning about 'capability'. *Animal Cognition, 12*, 267–286.
Webster, M. W., & Laland, K. N. (2017). Social information use and social learning in non-grouping fishes. *Behavioral Ecology, 28*, 1547–1552.
Wilson, E. O. (1975). *Sociobiology: The new synthesis.* Cambridge, MA: Harvard University Press.

Part II
Mainly of cats and dogs

Human history is marked by the variety of ways in which we interact with other animals. There are several types of domesticated animals with which large numbers of us share our daily lives and which give us a great deal of pleasure. We use animals for our entertainment: we stare at animals in circuses, zoos, aquaria, and dolphin parks; and for excitement we involve animals in sports such as horse and greyhound racing. There are other forms of entertainment involving animals including animals on the stage, and in cinema; we force animals to act in strange ways in advertising commercials to try to persuade us to purchase consumer items. It is a sad fact that cruelty to animals is prevalent in many parts of the world. At the most obvious, some people find enjoyment in harming animals by taking part in blood sports such as hunting, bear baiting, and hare coursing. Why do humans force animals into these roles? Are contemporary views changing the way we treat animals in these various contexts? Finally, for eons humans have survived by eating animals, a fact as true today as it was for the early humans. However, in some parts of the world the means by which we farm animals has changed radically and not always for the better. The way in which contemporary society reacts to mass farming takes a variety of forms ranging from changing the means of production, such as with free-range eggs, to changing our eating habits by not eating meat.

The following chapters consider what we know about these topics and the contribution psychology makes to their understanding.

2 Animals as companions

Our early ancestors saw animals as a resource which provided food, fur, and other materials that enabled them to survive in a harsh environment. The archaeological records are not definitive, but it is possible that the beginnings of the domestication of wild animals began about 300 centuries ago. The first domesticated animals, which gave food and other animal products, were probably goats and sheep followed by chickens. As civilisation advanced, so the human population increased and farming progressively became a means of food production. In keeping with this development, larger animals, such as horses and oxen, were domesticated to assist with tasks such as ploughing and transportation.

If the first domesticated animals served a utilitarian purpose, why did cats and dogs become so close to humans? There are various explanations for the beginnings of our lasting relationship with cats and dogs. Our relationship with the dog, a descendent of the wolf, has a long history, arguably stretching back over 30,000 years (Kotrschal, 2018; Vilà et al., 1997; Wang et al., 2016) and may have its origins in hunting where dogs were used to kill several different types of prey (Guagnin, Perri, & Petraglia, 2018). Over time, the dog's hunting skills were refined and the tamer breeds of dog developed abilities, such as retrieving fallen prey and herding other animals, that augmented the efficiency of human activities. The archaeological evidence that dogs were buried with their masters infers that humans and dogs had forged a strong psychological bond. The burial of dogs with people is found across ancient cultures from Siberia and Greece to China and Austria, a practice which speaks to the status, perhaps even a spiritual status, afforded to dogs (Morey, 2006). The ability of dogs to engage in a range of sophisticated cognitive tasks (Bensky, Gosling, & Sinn, 2013) and to respond to human facial emotional cues (Yong & Ruffman, 2016) no doubt enhanced their affinity with humans. The point has been reached, as described by Amiot and Bastian (2017), where the feeling of some people – primarily pet owners and vegetarians – towards animals is best expressed as *solidarity*.

The cat may have been welcomed by humans because it was useful in keeping down the numbers of rodents attacking grain stores. However, in some cultures the domestic cat was afforded a much more significant status. As Morey (2006) notes, the ancient Egyptians "Mummified cats in great numbers, and left the cats' remains in contexts that can be legitimately called cemeteries" (p. 168). Indeed, the ancient Egyptians bestowed high status on several animals, as seen by their ancient cemeteries which reveal the remains of birds, crocodiles, and gazelles. The cat may have been revered in cultures other than Egypt; in Cyprus there is evidence of cat burials which pre-date ancient Egypt (Vigne, Guilaine, Debue, Haye, & Gérard, 2004).

In today's world, our relationship with domesticated animals has taken several forms (Amiot, Bastian, & Martens, 2016). The most fundamental association arguably being with those animals we take into our homes and share our everyday lives. These animals take the role of a companion and, indeed, some of the literature uses the term *companion animals*. However, the term *pet* is more familiar and so is used interchangeably with *companion*.

It cannot be assumed that conceptualisations about pets are universal. Sevillano and Fiske (2016) make the point that we in the Western world hold stereotypes about certain animals; for example, the stereotype of the dog is of "man's best friend" leading to the view that dogs are friendly: as defines a stereotype, the dogs' friendliness extends to the majority of dogs. The stereotypical warmth we feel for man's best friend is not reciprocated in all parts of the world. Podberscek (2009) provides the example of South Korea where dogs have long been seen as a source of food and, indeed, eating dogs, sometimes for medicinal purposes, is seen as an aspect of cultural identity not to be interfered with or threatened by the West.

Pets and pet owners

How many people keep pets? What are the most popular pets? It is not an easy matter to estimate the size of the pet population and figures may vary significantly across studies. Murray, Browne, Roberts, Whitmarsh, and Gruffydd-Jones (2010) conducted a questionnaire survey of cat and dog ownership in 2,980 UK households. They estimate the UK cat population at 10,332,955 felines and 10,522,186 canines. In all, 26% of the households owned one or more cats while 31% of the households owned one or more dogs. In 2018, the company Statista reported a survey of pet ownership in UK households in 2017 and 2018. A substantial number of households had a pet (45%), with the dog the most popular: it was estimated that 26% of households owned a dog, suggesting a UK canine population of 8.5 million. The cat was the second most popular pet (18% of households), giving a feline population of 8 million. After cats there is a rapid fall to the rabbit (2%) in third place, followed by a long list of somewhat idiosyncratic choices including lizards (0.5%), rats (0.2%), and, least popular of all, mice (0.3%). The variations across surveys of pet ownership is due to several factors. The survey methodology may vary from questionnaires, to telephone polls, and personal interviews, which in turn may influence what people are prepared to report. There are shifts over time in pet ownership associated with economic factors and fashion.

Statista estimate that in 2017 the amount pet owners in the UK spent on veterinary and other pet services was over £4 billion. The estimated cost of keeping a dog, including food and insurance, is about £240 a month compared to about £100 a month for a cat. The large sums involved clearly show that pet ownership is a significant economic driver for a range of professions and retail outlets.

The Statista survey also revealed a range of reasons why people kept a pet: in answer to the question "What feelings do you experience as a result of owning a pet?", the three most popular answers were "Owing a pet makes me happy" (92% of respondents), "Owning a pet improves my life" (88%), and "Owning a pet is a privilege not a right" (88%). The survey confirms, of course, what the majority of pet owners know: a pet can bring a great deal of happiness and becomes an integral part of everyday household routines with feeding time, walks, and games. Indeed, pets can become part of a family, precipitating positive consequences such as the formation of strong attachments and increased levels of social interaction. The negative consequences can include child ill-health because of allergies (Paul & Serpell, 1966).

Given that dogs are the most common pet and, as Udell and Wynne (2008) suggest, the most "human-like" species, it is not surprising that the weight of research has focussed on the relationship between dogs and their owners. However, it is not a question of either a cat or a dog as a pet, everyday experience and empirical data show that cats and dogs can happily live together in the same household (Thomson, Hall, & Mills, 2018). While the owner–cat relationship is not as thoroughly researched as that of the owner–dog relationship, there are similarities as well as differences between the two (Pongrácz & Szapu, 2018).

Feline companions

A stereotype of the cat is that it is a somewhat aloof, solitary animal, altogether less sociable than the dog: we joke that "Dogs come when they're called; cats take a message and get back to you later" or, as Kirk (2019) puts it, "Dogs have masters, cats have staff." Bradshaw (2016) explains how this view of cats is a product of history: dogs have been domesticated for several millennia longer than cats, allowing them time to evolve to become much more socially interactive with humans than cats. Of course, cats are now commonplace as companion animals and bring a familiar mixture of pleasure and tribulations to their owners (Bernstein, 2007). Turner (2017) highlights several areas of interest in feline–human interaction including cat–human communication, cat–owner personalities, and problems caused by cats.

Cat–owner interactions

What is it that constitutes a satisfying cat–owner relationship from the owner's perspective? Howell et al. (2017) developed the 33-item Cat-Owner Relationship Scale (CORS) to assess owners' views of their relationship with their cat. They report that this scale contained the three subscales of *Pet-Owner Interactions*, *Perceived Emotional Closeness*, and *Perceived Costs*. Sample items for these subscales are shown in Table 2.1.

The items comprising the Pet-Owner Interactions subscale of the CORS indicate the nature of the activities that form interactions between cat and owner. Thus, owners who regularly interact with their cat spend time playing games with their cat, they will stroke and pet it, talk to it, watching its actions and have it close by when relaxing. Pongrácz

Table 2.1 Sample Items from CORS (Howell et al., 2017)

Pet–owner interactions
How often do you play games with your pet?
How often do you cuddle your pet?
Perceived emotional closeness
My pet gives me a reason to get up in the morning.
My pet provides me with constant companionship.
Perceived costs
It bothers me that my pet stops me from doing things I enjoyed before I owned it.
It is annoying that sometimes I have to change my plans because of my pet.

and Szapu (2018) surveyed a sample of Hungarian cat owners asking about their relationship with their pet and found many similarities with dog owners. However, cat owners were strongly of the view that their cat is a family member with a high level of socio-cognitive understanding of human emotion and nonverbal behaviour. Given these views, cat owners used nonverbal behaviour, particularly pointing and visual cues such as gazing, to communicate with their cat.

Arahori et al. (2017) compared Japanese cat and dog owners' views relationship with their pet and their views of their pets' emotions and intellect. While cat and dog owners frequently saw their pets as a family member, perhaps in contrast to Pongrácz and Szapu this view is stronger for dog owners. In addition, dog owners were more likely to attribute emotional and intellectual abilities to their pets.

The contrasting findings of the Hungarian and Japanese studies maybe accounted for in three, not mutually exclusive, ways. There may be sampling differences in age, gender, and so on; the studies used different questionnaires; and the variation in findings may reflect genuine cultural differences. The development of standardised survey instruments will assist in future research so allowing greater confidence to be ascribed to genuine cultural differences as a cause of variation (Duffy, de Moura, & Serpell, 2017).

Cat–owner personalities

Bennett, Rutter, Woodhead, and Howell (2017) assembled a list of over 200 adjectives that could potentially be used to describe a cat's personality. Two focus groups then slimmed down this list to 118 words. In the next part of the study 416 adult cat owners rated a familiar cat on each of the 118 words. The analysis of the ratings yielded the six personality dimensions of Playfulness, Nervousness, Amiability, Dominance, Demandingness, and Gullibility. As well as a research instrument, Bennett et al. suggest that the six personality factors could be put to use by cat adoption programmes in matching cats with prospective owners. This suggestion is reinforced by the finding of Finka, Ward, Farnworth, and Mills (2019) that an owner's personality is related to their cat's behaviour, welfare, and lifestyle.

Gosling, Sandy, and Potter (2010) compared the personalities of 4,565 participants, divided into the four groups of self-identified dog person, cat person, both, or neither, using the self-report version of the Big Five Inventory (BFI; John, Naumann, & Soto, 2008). The Big Five personality dimensions are Agreeableness, Conscientiousness, Extraversion, Neuroticism, and Openness. Gosling et al. found that the dog people scored higher than cat people on Agreeableness, Conscientiousness, and Extraversion, and lower on Neuroticism and Openness: these differences remained when sex differences in pet–ownership rates were controlled. With the exception of Neuroticism, where they scored highest, the cat people group tended to be lower than the other three groups on the remaining four dimensions.

Gosling et al. make the suggestion, in keeping with Bennett et al. (2017), that their findings could be put to practical use: "Self-identification as a certain type of pet person may also provide relevant and practical information for areas such as pet selection within animal shelters, pet welfare, and other human–animal relationships. Pet person identifications could also be useful in healthcare settings (e.g., hospitals, mental healthcare facilities, nursing homes), where an affinity for certain types of animals may affect the selection of species used in pet therapy" (p. 221).

A study by Evans, Lyons, Brewer, and Tucci (2019) supports the suggestion that there may be benefits to taking account of personality in an exercise matching 126 cats and owners. They found that owners expressed greater satisfaction with cats high in agreeableness and low in neuroticism (see the "Feline Five" below). The owner's impulsivity and the cat's agreeableness correlated with higher satisfaction, as did a contrast in owner dominance and cat agreeableness.

Litchfield et al. (2017) carried out a study in Australia and New Zealand to explore the personalities of 2,802 pet cats. A sample of owners completed a survey, rating their cats on 52 personality traits gathered from previous studies. The analysis revealed the "Feline Five" personality factors of Agreeableness, Dominance, Extraversion, Impulsiveness, and Neuroticism. Litchfield et al. suggest that knowledge of the factors could be used to improve cat welfare; for example, highly impulsive cats could react easily to environmental stressors.

Problems caused by cats

Turner (2017) lists the problem caused by cats as "Allergies, bites and scratches on owners and non-owners, zoonotic diseases, and predation" (p. 302). The issue of predation is considered in detail in Chapter 9, leaving health matters and aggression.

Health matters

There are vaccines available for the treatment of allergies, which can cause skin problems and breathing difficulties, although the simple solution for those who are strongly allergic to cats is to find another companion animal. Zoonotic diseases are brought about by bacteria, parasites, and viruses which cross between animals and humans (Murugan et al., 2015). These diseases can be serious, such as with the *Ebola* virus and salmonellosis, or more manageable as with "cat scratch disease," a bacterial infection of an open wound caused by a scratch or bite. The risk of ill-health can be managed close to home, as with other pets, by a good health-care regime for the cat including regular vaccinations. On a larger scale, coordinated initiatives such as instigating and maintaining comprehensive records and standardised education for professionals working with animals may bring widespread benefits (Sterneberg-Van der Maaten, Turner, Van Tilburg, & Vaarten, 2016).

A cat which is not house trained may be a health risk through soiling either by indoor elimination of urine and faeces (Barcelos, McPeake, Affenzeller, & Mills, 2018; Heath, 2019) or urine spraying, perhaps to mark territory (Horwitz, 2019). The cause and management of these problems is discussed in Chapter 5.

Aggression

Feline aggression can take a variety of forms serving different purposes in different environments; for example, it may be offensive or defensive, predatory, a form of play, territorial, or a consequence of stress or fear (for a succinct summary, see Penar & Klocek, 2018). However, for owners of domestic cats, their cat's aggression becomes a problem when it is directed at them personally.

A Spanish study reported by Palacio, León-Artozqui, Pastor-Villalba, Carrera-Martín, and García-Belenguer (2007) looked at animal aggression towards people in the region of Valencia between 1995 and 2000. They found a total of 12,040 recorded acts of animal aggression towards people, of which 89% involved dogs, 8% cats, and 3% other species including horses, monkeys, and rodents. For felines specifically, there was an average incidence of 6.36 aggressive acts per 100,000 people per year: the average incidence was greater for women (7.1 acts of aggression per 100,000 people per year) than men (4.6), and greater for children aged from 0 to 14 years (6.8) than for people aged from 15 to 64 years (5.1) and those over 65 years of age (4.7).

Palacio et al. (2007) looked at the nature of the bite wounds by cats that were mainly single punctures on the hands. In children, the head and neck areas were bitten more than for adults. The cats involved were mostly unowned; female Siamese cats were prevalent in cats that attacked their owners. The most common situation of a bite was a defensive response to a threat. The most serious bites, requiring medical assistance, were from unowned cats.

Amat and Manteca (2019) note that owner- and family-directed aggression is common in cats, particularly in single-cat households. They describe the risk factors for this type of aggression as obtaining the cats from pet shops, poor early socialisation with people, and if the cat is not allowed outdoors. While not as extreme as aggression, destructive scratching of household items such as carpets, furniture and window frames can be both aggravating and expensive for the owner (DePorter & Elzerman, 2019). The effect of aggression, not surprisingly, is to increase the likelihood of euthanasia or the cat being put into a shelter.

The nature of the owner–animal relationship

The nature of the relationship and the positive attachments between owners and their companion animals has, like any other relationship, an array of dimensions which may be approached from various theoretical standpoints ranging from the biological to the psychological to the cultural (Beck, 2014; Echeverria, Karp, Naidoo, Zhao, & Chan, 2018; Herzog, 2014; Hosey & Melfi, 2014; Odendaal & Meintjes, 2003). The research exploring the owner–animal relationship is, like many other areas of psychology enquiry, sensitive to the methodology employed. In addition, the use of the theoretical terminology in this context should be qualified as exemplified by the concept of *attachment*. As discussed above, there is attachment theory in the context of human–human relationships (e.g., Bowlby, 1953, 1956) and while we may think of human–animal attachment as having similar qualities it does not follow that the two are identical (Crawford, Worsham, & Swinehart, 2006). Thus, Rehn, Lindholm, Keeling, and Forkman (2014) demonstrated that while owners form an attachment with their dog, as with other companion animals, a reciprocal relationship cannot be proven. The dog *may* form an attachment but its behaviour towards its owner may equally well be seen as a consequence of positive reinforcement.

Nonetheless, at its most obvious, the relationship with their pet can bring the owner the personal and social rewards afforded by daily companionship: these rewards may vary according to the type of animal (Zasloff, 1996), or the animal's behaviour, or the owner's preferences. Blouin (2013) conducted interviews with 28 dog owners in the midwestern United States, finding that the owners had three distinct orientations towards their pets. The first orientation Blouin labelled *dominionistic*: this type of owner has a utilitarian view of their pet, seeing them as an object only to be valued for their usefulness as, say, a guard dog. Those owners with a *humanistic* outlook approached their pet as a surrogate human and took pleasure in a close emotional attachment with their dog. Finally, a *protectionistic* orientation was marked by a high regard for all animals, including pets as important companions but also as animals in their own right.

Statts, Sears, and Pierfelice (2006) asked a sample of 302 American male and female students about pet ownership. The students gave five main reasons for keeping a pet: (i) to keep active (21.5%); (ii) to prevent loneliness (18.2%); (iii) the pet serves a useful function (14.2%); (iv) keeping the pet for someone else (12.9%); (v) the pet helps when times are hard times (10.6%). There was a gender difference such that the women were more likely keep pets for social support, to help through hard times, or to combat loneliness. Men were more likely own pets in order to keep active, or because the pet served a useful function, or to look after the pet for someone else.

The dog owners questioned by Maharaj and Haney (2015) provided a succinct summary of what a dog brings to its owner's life. The first point reflects the interpersonal relationships involving the dog playing a role in family routines. The second point lies in the owner's experience of their dog as a subjective being with its own personality and ability to communicate and understand the owner. Finally, there are the psychological and health benefits associated with dog ownership. It is important to state that careful attention to puppy training is crucial to the future happiness of both owner and dog (González-Martínez et al., 2019).

However, alongside these commonplace rewards there are other dimensions to consider: Tipper (2011a) reflects that pets may serve the function of reinforcing the owner's social and psychological identity:

> Unusual pets such as rats or snakes might be part of an "alternative" identity; cats, poodles, or Chihuahuas might be valued for their "feminine" associations, whereas others may seek the "masculine" image of dogs such as Alsatians; and British bulldogs, Yorkshire, or Scottish terriers may express a regional or national identity. (p. 87)

When it comes to the choice of a pet not only are there a range of potential species but also choice of breed within a species. The decision as which dog to take as a pet can be influenced by practicalities, such as size age or sex (Boruta, Kurek, & Lewandowska, 2016), or what is fashionable at the time. Ghirlanda, Acerbi, Herzog, and Serpell (2013) looked at the popularity of different breeds of dog in America between 1926 and 2005. There was no evidence that those breeds having more desirable behaviours, such as ease of training and being longer lived or with fewer inherited genetic disorders were more popular than other breeds. Ghirlanda et al. conclude that a breed's popularity at a given time is not so much due to its intrinsic features but how fashionable it seen to be. An example of a fashionable breed is found with the rise in popularity of the Chihuahua at a given time as influenced by Paris Hilton and her favourite dog, Tinkerbell Hilton (Redmalm, 2014). Some dog owners remain immune to fashion and stay loyal to a particular breed of dog because they like its temperament or appearance (Sandøe et al., 2017).

In addition, Tipper (2011a) suggests that for many people who experience the ache of loneliness a pet can be a substitute for people. A companion animal becomes someone to share one's life with: we offer care and in return receive attention and, to borrow Tipper's (2013) vivid phrase, perhaps existential moments of being. Evans-Wilday, Hall, Hogue, and Mills (2018) explored how owners may disclose difficult personal facts to their dogs as opposed to their partner or a confidant such as a close friend. They found that: "Dog owners reported greater willingness to talk to their dog (and partner) compared with a confidant across the emotional disclosure topics of depression, jealousy, anxiety, calmness, apathy, and fear. For topics relating to jealousy and apathy, dog owners showed greater willingness to talk to their dog than their partner and confidant" (pp. 361–362). Evans-Wilday et al. conclude that dogs can play a similar role as partners in disclosure of emotions although this does not preclude talking to partners and confidants. An example, perhaps, of the dog as our best friend.

Nast (2006) describes how pets can become a fashion statement, typified by celebrities prepared to spend outrageous sums of money on pet accessories. Nast notes that a doggie boutique in Los Angeles, *Fifi & Romeo* which is frequented by Hollywood stars, sells miniature cashmere sweaters for $200 and raincoats for $105. The dogs of the rich and famous in LA are also well catered for at canine spas such as *Dog House* where they can indulge in massages and herbal wraps. The exaggeration of the emotional bond between owner and pet, which as Nast notes can over-step standards of propriety, is used as a means of justifying commercialisation and scandalous levels of expenditure on animals (Vänskä, 2016).

A related phenomenon lies in adults without children, whether through choice or not, treating pets as substitute children. The psychological boundary between pets and children become blurred to the point of coining names such as "fids" (feathered kids) for parrots (Anderson, 2003, 2014) and "fur babies" for dogs, treated at "Yappy Hour" at a bakery cooking delicacies for dogs (Greenebaum, 2004). This view of pets as intimate companions is a two-edged sword: on one hand it may lead to high levels of care; on the other hand, it may produce a poor social, dietary, and exercise regime for the animal.

Many pet owners will either buy their pet from a commercial outlet, a pet shop, or professional breeder, or give a home to a rescue dog from an animal charity. The choice of where to obtain a pet is not a neutral action, the location selected may have a profound effect on the eventual relationship between owner and animal.

"Puppy farms" and pet shops

The high-volume breeding of dogs in Commercial Breeding Establishments (CBE), sometimes referred to as "puppy farms" or "puppy mills", has raised concern about the short- and long-term prospects for the puppies and their mothers in such establishments (McMillan, Duffy, & Serpell, 2011). If the retailer does not have high standards of hygiene, particularly in cases of "high-volume" breeding, there is an elevated risk of disease (e.g., Schumaker et al., 2012). It is evident that, as with many other species including humans, the dog's experiences during the first year of life have a formulative influence on its later temperament and behaviour (Foyer, Bjällerhag, Wilsson, & Jensen, 2014). McMillan (2017) compared owners' reports about dogs from CBEs with similar reports about dogs obtained from other, primarily non-commercial, sources. McMillan stated that the data, drawn from studies in the UK, Australia, Italy, and the United States:

> Suggest that dogs sold through pet stores and/or born in high-volume CBEs have an increased frequency of a variety of undesirable adulthood behaviors compared with dogs from other sources, particularly noncommercial breeders. The most common finding (6 of 7 reports, or 86%) was an increase in aggression directed toward the dog's owners and family members, unfamiliar people (strangers), and other dogs. The most consistent type of increased aggression .... was aggression toward owners and family members. The other characteristic found in multiple studies was increased fear .... which was in response to strangers, children, other dogs, nonsocial stimuli, and being taken on walks. (p. 24)

As McMillan notes, these data are based on owners' reports and so are subject to verification from other sources. In addition, not all commercial breeders should be tarred with the same brush, some breeders have exemplary standards. Gray, Butler, Douglas, and Serpell (2016) compared Pug, Jack Russell, and Chihuahua adult dogs raised from puppies bought from responsible breeders with those acquired from less responsible breeders. The dogs from responsible breeders were better adjusted across several dimensions such as aggressive behaviour and fear of other dogs. Gray et al. suggest this finding demonstrates the importance of owners acquiring puppies from breeders who follow the appropriate guidelines of organisations such as the Royal Society for the Prevention of Cruelty to Animals (RSPCA) and the British Veterinary Association (BVA).

McMillan, Serpell, Duffy, Masaoud, and Dohoo (2013) used owner reports of their adult dog's behaviour to compare outcomes for puppies purchased from pet stores with puppies from noncommercial breeders. The dogs obtained as puppies from noncommercial breeders had fewer problems with aggression, fear of other dogs, and housetraining. McMillan et al. suggest that as compared to noncommercial breeders, dogs from pet stores have a greater risk of developing undesirable behaviours.

Animal shelters

There are numerous animal charities, such as the RSPCA in the UK, which has almost 50 shelters, and the American Society for Prevention of Cruelty of Animals (ASPCA), that give a home to stray or unwanted animals. Animals may need shelter because of changes in the owner's life such as bereavement, loss of employment, or a change in family composition such as a newborn child. A survey of cat shelters in Sweden by Eriksson, Loberg, and Andersson (2009) found that the three most common reasons for relinquishing a cat were allergy, moving house, and that the cat was homeless. Diesel, Brodbelt, and Pfeiffer (2010) looked at why owners decide to relinquish their dogs to an animal shelter. The study took place at 14 shelters in the UK with a sample of 2,806 dogs. They found that the two most frequent explanations for relinquishment were the dog's problem behaviour, including aggression and destructiveness, and that the dog needed more attention than they had time to give. In a substantial number of the cases, the relinquished dogs had been obtained with little or no planning or advice. Diesel et al. suggest that levels of relinquishment due to owner-related problems could be ameliorated by providing advice about the dog when ownership is taken and by monitoring the progress of adopted dogs. The dog's problem behaviours can be tackled through dog training classes for adopters.

In countries larger than the UK, the issue is magnified. In the United States, there are approximately 13,600 animal shelters with an annual population of approximately 7.6 million cats and dogs. An estimated 31 to 55% of these animals are put down each year. In Taiwan, there are cultural and religious traditions that oppose the killing of unwanted animals and which act to encourage abandoning dogs. This situation has led to a huge increase in the numbers of stray dogs (Hsu, Severinghaus, & Serpell, 2003). In a similar vein, some countries have a "no-kill" policy for animal shelters; for example, since 1981 in Italy it is illegal to put down unwanted stray dogs unless they are either dangerous or terminally ill. However, as Dalla Villa et al. (2008) explain, the subsequent rise in the number of stray dogs in Italy has meant that this policy has had the undesired effect of warehousing dogs in dubious conditions for the remainder of their lives.

A large number of animals in the same place is stressful and increases the risk of disease, resulting in an environment liable to have a detrimental effect on the animal's welfare and behaviour. This unfavourable situation creates a double-bind such that the stressful consequences following being rescued act to lower the chances of the animal's adoption by a new owner. It is important therefore to ensure that the shelter provides a healthy environment and prevents deterioration in the animals' behaviour (Cozzi, Mariti, Ogi, Sighieri, & Gazzano, 2016; Stella & Croney, 2016). The application of veterinary measures to improve the kennels, such as regular behavioural and physical examinations, can bring about marked improvements in the animals' health (Dalla Villa et al., 2008). However, given the wide range of needs involved in animal welfare – encompassing health and veterinary issues, housing and environment, and breeding and reproduction (Rioja-Lang, Bacon, Connor, & Dwyer, 2019) – there is a great deal for shelters to attend to.

Alongside animal welfare, animal shelters aim to rehome the animals. It makes sense that visitors to a sanctuary who are contemplating an adoption see the animals at their best. There are two points for shelters to consider in seeking to maximise their chances of rehoming an animal. First, the presentation of the animals as healthy and free from disease and other problems. Second, attention to the specific factors that may influence the decision making of prospective adopters who will come with their own expectations of what dog ownership will mean for them (Powell et al., 2018).

The first point requires an efficient veterinary programme to ensure the animals are in good health (e.g., Dalla Villa et al., 2008), particularly so for older cats and dogs (Hawes, Kerrigan, & Morris, 2018). The second point, prospective adopter decision making, can be looked at empirically in two ways: (i) which animals are *not* chosen by adopters and so stay longest in the shelter; (ii) what criteria do prospective adopters use when making a positive decision to adopt?

Adopter decision making

The process of decision making can be a complex psychological process involving person variables, such as previous experience and emotional state; and environmental factors such as the amount information available about the animals (Newell, Lagnado, & Shanks, 2015). There are several variables, including the animal's age, breed, physical size, length of coat, and sex, which may influence adopter decision making and thereby determine an animal's length of stay in a shelter (Brown, Davidson, & Zuefle, 2013; Cannas, Rampini, Levi, & Costa, 2014; Protopopova, Gilmour, Weiss, Shen, & Wynne, 2012; Protopopova & Wynne, 2016; Žák, Voslářová, Večerek, & Bedáňová, 2015). As may be anticipated, type of breed is an important factor for prospective adopters with sporting and fighting breeds the least attractive choices; the ratters, lap, and toy breeds stay shortest before adoption. It may be that some breeds of dog, such as Pit Bull terriers (see the later discussion of the Dangerous Dogs Act 1991), have such negative stereotypes that they are less attractive to prospective adopters (see Wright, Smith, Daniel, & Adkins, 2007). It follows that the way certain breeds of dog are labelled and described within the shelter can influence adopter perceptions (Gunter, Barber, & Wynne, 2016). For smaller animals, viewing at eye level is an important consideration in attracting visitor attention (Fantuzzi, Miller, & Weiss, 2010).

Protopopova, Gilmour, Weiss, Shen, and Wynne (2012) found that dogs given up by their owners, as opposed to strays, had a shorter length of stay in the shelter. The dog's breed and how it came to arrive at the shelter are *static* variables in that they cannot be changed. However, there may be *dynamic* variables which can be changed to increase the chance of adoption.

Protopopova, Mehrkam, Boggess, and Wynne (2014) found that when controlling for appearance, three specific behaviors had a significant effect on a dog's length of stay: (i) if it leant or rubbed on the kennel wall; (ii) if it faced away from the front of the enclosure; and (iii) if it stood rather than sitting or lying prone. If these negatively perceived behaviours can be changed, then it may be that the chances of adoption will increase. To this end, several environmental enrichment and behaviour change programmes have been developed (e.g., Demirbas et al., 2017; Protopopova, Hauser, Goldman, & Wynne, 2018; Protopopova & Wynne, 2015).

What do visitors actually do when they enter a rescue sanctuary and how is this related to their decision regarding adoption? Protopopova et al. (2012) found support for the adage that first impressions count: "Adopters were likely most influenced by variables that were readily observable in a few seconds, such as the overall look of the dog and the information that was written on the cage card" (p. 68). Wells and Hepper (2001) observed the behaviour of 76 visitors to a rescue shelter for dogs in Northern Ireland. The average visitor paused to look briefly at just less than one-third of the dogs, mostly those close to the entrance to the kennels, and just 3 of the 76 actually purchased a dog. An American study by Garrison and Weiss (2015) found that prospective owners behaved like consumers: they valued variety in the available dogs and were willing to travel to find a dog that matched their preferences. Garrison and Weiss suggest that "A comprehensive animal relocation program that transports a variety of dogs, not just puppies or small dogs, and that is well marketed to the public has the potential to significantly increase traffic and therefore adoptions at animal welfare organizations" (p. 69). In a similar consumerist vein, Reese, Skidmore, Dyar, and Rosebrook (2017) suggest that American shelters should vary their prices according to the popularity of certain characteristics; thus, puppies would cost more than older dogs and pedigrees more than mongrels. This consumerist view of pets at commodities and fashion accessories may be part of an explanation for the large numbers of animals in sanctuaries. As fashions change and the realities of having to care for an animal are felt, so the commodity is disposed of to make way for whatever fad follows. So the cycle continues.

Not all adoptions are successful and some dogs are returned to the shelter after adoption, although a return does not necessarily preclude further adoptions. Patronek and Crowe (2018) noted that of the 816 shelter dogs returned from adoption over a 2-year period, 695 were subsequently re-adopted. As shown by studies from several countries, there is a range of factors which influence an adopter's decision to return an animal including its aggression, fearfulness, destructiveness, excessive barking, anxiety, and it not adapting to children or other pets (e.g., Chung, Park, Kwon, & Yeon, 2016; Gates, Zito, Thomas, & Dale, 2018; Normando et al., 2006; Shore, 2005; Vitulová, Voslářová, Večerek, & Bedáňová, 2018). Adoption is not the only strategy that shelters can use to give animals a better quality of life. Australian studies by Kerr, Rand, Morton, Reid, and Paterson (2018) and by Rand, Lancaster, Inwood, Cluderay, and Marston (2018) used various strategies, such as fostering and liaison with other rescue groups, successfully to increase rehoming and reduce levels of euthanasia in sheltered cats and dogs.

Some owners may decide to have more than one dog which is advantageous if the dogs are left alone but it is uncertain whether the dogs form relationships with each other

(Mariti, Carlone, Ricci, Sighieri, & Gazzano, 2014). Once a pet is acquired and assimilated into household routines so the relationships, activities and routines of members of the household change.

Walking the dog

One of the many changes in routine associated with a new pet lies in establishing an exercise regime. While some animals, notably cats, take care of their own routines, others such as dogs and horses need regular exercise to maintain their health and fitness. For many dog owners the daily activity of taking the dog for a walk becomes part of their life. While we may think of walking as a functional activity by which we move from one location to another, walking can serve many varied purposes: walking can be purposeful, or for pleasure, or a means to fitness; walking may be an aimless wander or a purposeful march; walking provides a way to be close to different landscapes; we may walk alone or as part of a group; and walking can be competitive over short or extremely long distances (Edensor, 2000).

The routine of walking the dog can incorporate any or all of these aspects; however, with regard to dog walking, two aspects have received particular attention. First, the social dimension where, particularly in suburban areas, the morning or evening dog walk becomes a catalyst for social interactions between walkers (McNicholas & Collis, 2000) and, of course, their dogs. Second, the health benefits associated with taking the dog for a walk (Westgarth et al., 2019).

Dog walking as a social activity

Fletcher and Platt (2018) follow the spirit of Edensor's (2000) critique of walking in considering the more profound aspects of dog walking. They suggest, for example, that within the seemingly simple act of taking the dog for a walk lies the potential for a fuller understanding of animal–human interactions. A series of interviews with dog walkers revealed a range of subplots. Some owners said that the walk was for the dog's benefit, to give it exercise and help it keep fit and healthy and saw this activity as part of their personal obligation to the animal. These owners would try to take their dog to the appropriate type of setting for their breed so that they could be let off their leash to do what comes naturally such as chasing rabbits, rummaging around in the undergrowth, and interacting with other dogs (Řezáč, Viziová, Dobešová, Havlíček, & Pospíšilová, 2011). Fletcher and Platt note that one owner, Jane, said that by taking regular walks she felt she was righting a wrong that had been inflicted on their rescue dog, Copper: "Jane believed that, as Copper's early life was subject to human neglect, it was now her human family's responsibility to ensure his life with them was filled with love and enjoyment" (p. 219).

The dog walkers were aware of the people around them so that they knew other dog walkers and which of them would stop and chat or simply walk on with a wave of acknowledgement. The walkers were aware that other people would be walking who did not have a dog and who may dislike or be afraid of dogs. The dilemma which then arises is whether to allow the dog to go off-leash so it can roam freely but at the potential cost of frightening other walkers. The walker's decision to allow a dog off-leash may be influenced by the setting and age and type of dog (Sediva, Holcova, Pillerova, Koru, & Řezáč, 2017). In all, the act of dog walking is like many other areas of human social activity: there are positive aspects, such as building social networks, and negative features such as the formation of cliques and outgroups (Graham & Glover, 2014). Yet further, the act of dog walking provides an example of how the affiliation between owner and dog can lead to the synchronisation of their physical activity. A French study reported by Duranton, Bedossa, and Gaunet (2018) recorded the pace of freely walking dogs (i.e., off the leash) when their owners walked quickly, walked slowly, or stood still. They found that the dogs systematically changed their walking pace to match that of their owner. Duranton, Bedossa, and Gaunet suggest that the dog's behaviour may be a result of their natural inclination to follow a leader or the product of an everyday experience (or both).

The presence of large numbers of dogs in a city's open public space can bring problems such as aggression and dog waste. The need to exert some regulatory control over dogs in urban settings has become a pressing issue in some countries (Carter, 2016).

Dog parks

The issues raised by off-leash dog walking have been addressed by the creation of dog parks (Instone & Sweeny, 2014; Weston et al., 2014). A dog park is a fenced-off open area, typically within a large city, where owners are allowed to let their dogs off the leash. The parks may provide facilities such as drinking water and dog waste disposal amenities.

Instone and Mee (2011) describe how in Australia tensions between dog owners and non-dog owners about dogs in public spaces became a legislative matter. The position was reached where local governments were required to provide off-leash spaces to compensate dog owners for restrictions on dogs in urban locations. The designation of a city area of open ground as a dog park is itself not without problems and can bring about local negotiation and tensions (Rock et al., 2016). Rahim, Barrios, McKee, McLaws, and Kosatsky (2018) reviewed the literature on dog parks, summarising their costs and benefits. The costs include financing the management of public hygiene and safety; the benefits include increasing social connectedness alongside the health benefits of walking for the dogs as well as the owners. In the UK, there are parks with fenced-off areas, in some cases called *dog paddocks*, specifically for dog walking and training (see www.dogwalkingfields.co.uk).

The health benefits of walking the dog

We are all extolled to walk more both for our own fitness and health and to reduce the pollution from travelling short distances in cars. It is not surprising that dog owners, both male and female, spend more time walking each week than those who do not own a dog (Brown & Rhodes, 2006; Cutt, Giles-Corti, Knuiman, Timperio, & Bull, 2008). However, it appears that those owners who potentially have the most to gain are the least likely to walk their dog (Coleman et al., 2008). Yet further, some owners construe the benefits of dog walking primarily in terms of meeting the dog's needs rather than their own (Westgarth, Christley, Marvin, & Perkins, 2017).

Christian et al. (2016) review the evidence that supports the view that dog walking has health benefits and suggest a range of strategies – providing health advice, the formation of dog walking groups, school or workplace activities, and media campaigns – to promote walking the dog. Christian et al. note that promoting dog walking would mean that "Advocacy for dog walking–oriented policy relevant initiatives are needed, starting with park development, dogs-allowed policies, off-leash zones, and dog-friendly built environments" (p. 240). The enactment of such policies would come at a price, raising the perennial question of just who pays the bills.

Young pet owners

Animals can play rich and varied roles in a child's life, see Jalongo (2018) for an overview, including the role of companion animal. Marsa-Sambola et al. (2016a) reported a survey of pet ownership amongst over 14,000 female and male 11–15-year-olds in England, Scotland, and Wales. They found that in keeping with studies carried out in Australia, Germany, and the UK, a large number of adolescents (72% of their sample) lived in households with a pet. The older adolescents, both female and male, were more likely to own a dog, while the younger adolescents were more likely to own amphibians, fish, or reptiles. As noted by Purewal et al. (2017), there are a range of emotional health benefits, including higher self-esteem and less loneliness, seen in those adolescents who keep pets.

A child taking responsibility for a pet highlights two psychological issues in the child's development: (i) learning to take responsibility for the animal's welfare; (ii) forming an attachment with a pet.

Taking responsibility

Muldoon, Williams, and Lawrence (2016) note that taking responsibility for an animal can promote the child's concern for animals generally as well as helping to develop an ethical and moral sense of caring. They conducted a series of focus groups discussing animals with Scottish children aged from 7 years to 13 years. There was a wide range of knowledge about animals in general, alongside examples of specific familiarity in that some children knew what some animals, typically cats and dogs, preferred to eat but were unsure about other pets such as hamsters. Some children showed a high level of knowledge about fishkeeping, understanding the importance of both creating the right aquatic environment and proper feeding. Muldoon, Williams, and Lawrence suggest that this knowledge may be a consequence of previously losing fish and learning from their errors.

The children said that some animals, such as dogs, had feelings and had the status of a friend capable of shows of affection. The child's perception of the animal's behaviour and internal state was made with reference to their own functioning. While this strategy may work well in certain instances, such as recognising when their pet is hungry, it is altogether less effective in other situations. Lakestani, Donaldson, and Waran (2014) reported that young children, aged from 4 to 6 years, were poor at recognising certain animal states, particularly fear, because they focussed on the wrong cues. A dog's fear most accurately perceived by attending to its posture and tail position rather than its face.

Forming attachments

Tipper (2011b) considered the nature of the relationship between children and animals by drawing on data from interviews with 49 children, 31 girls and 18 boys, from a range of backgrounds living in the north of England. Tipper's analysis showed how children regard pets as part of a family: this applies not just to their own family but also to relatives and friends. In everyday life, the children had their ups and downs with the family pet, describing fun times and quarrels as with any family member. Charles (2014) reinforces Tipper's findings, charting the history of pets as a loved family member. Indeed, should there be any doubt that pets are a fundamental member of a family, there are legal cases, as with child custody, as to who gets the pets when the family breaks up (Rook, 2014) and the legal standing of animals in the housing arena (Rook, 2018). Yet further, in natural disasters people will risk their lives, as they might do for a relative, rather than leave their pets to their fate (Irvine, 2009).

Marsa-Sambola et al. (2016b) constructed a scale, *The Short Attachment to Pets Scale (SAPS)*, to assess a child's attachment to pets. SAPS is intended for use in research into positive emotional bonds with pets and in the evaluation of interventions intended to promote positive relationships with animals. The nine items comprising SAPS are based on material drawn from the Health Behaviour in School-Aged Children Survey conducted in England and Scotland with school pupils aged 11, 13, and 15 years old. SAPS uses a 5-point response scale from 1 (strongly agree) to 5 (strongly disagree) with respondents self-rating on items such as "I have sometimes talked to my pet and understood what it was trying to tell me," "I consider my pet to be a friend," and "My pet knows when I'm upset and tries to comfort me." As a generalisation, younger children show a greater attachment to pets than older children as do females compared to males.

Hawkins, Williams, and Scottish Society for the Prevention of Cruelty to Animals (Scottish SPCA) (2017) conducted a self-report survey in Scotland of attachment to pets, using the SAPS scale (Marsa-Sambola et al., 2016b) with 1,217 male and female 7- to 12-year-old primary school children. The children, girls more so than boys, had close attachments with their pets, particularly to cats and dogs but also to horses and small mammals. The level of attachment was higher when the child saw the animal as their own rather than a family pet. Attachment was also strongly related to caring behaviours, such as time spent cuddling, stroking, and playing with the pet; and in friendship behaviours such as talking to pets, including sharing secrets, and crying when sad.

Hawkins et al. also reported that an attachment to pets is a significant predictor of a positive attitude to animals in general. This positive attitude is manifest in a concern for the humane treatment of animals and advocating less animal cruelty (Paul & Serpell, 1993; Thompson & Gullone, 2008). It follows that if children can be encouraged to care for a pet, there may be both short- and long-term beneficial outcomes for both the children and the animals. In addition, when their children walk a dog this can encourage parental interest in aspects of safety, such as traffic speed, in the immediate environment (Roberts, Rodkey, Grisham, & Ray, 2017).

Westgarth et al. (2013) reported that of a sample of primary school children, aged 9–10 years, those with a dog were likely to take it for a walk at least once daily and often more. In a time of public health concern about childhood obesity, the possibilities for exercise offered by dog walking have been noted in several countries including America (Gadomski, Scribani, Krupa, & Jenkins, 2017), Australia (Christian, Trapp, Lauritsen, Wright, & Giles-Corti, 2013) and the UK (Westgarth et al., 2017). Gadomski et al. make the point that a strong attachment between a child and their pet enhances the likelihood of a successful exercise routine becoming established.

Hirschenhauser, Meichel, Schmalzer, and Beetz (2017) conducted a questionnaire study of attachment to pets among two groups of children, one aged 6–10 years, the other 11–14 years in the Austrian city of Linz. The children had a range of pets from the commonplace cats and dogs, through smaller animals such as rabbits and mice, and to reptiles and fish. Hirschenhauser et al. reported that the 6–10-year age group described closer attachments to cats and dogs than to birds or reptiles; this relationship between level of attachment and the taxonomic order of the pets was not evident in the 11- to 14-year-olds. In the younger group, there was no association between attachment to a pet and gender; for the older children, the girls showed greater attachment than the boys. Hirschenhauser et al. suggest that the age effect on attachment to a pet may be explained by the younger children's continued development of an understanding of animals; on the other hand, the older children may spend less time with their pets due to competing demands.

A gender effect with girls showing a higher attachment to their pet was evident for the older but not the younger children. Hirschenhauser et al. do not discount the possibility of a sampling effect in their study but note that this finding is in keeping with previous research (Herzog, 2007; Kellert & Berry, 1987; Phillips et al., 2011). The gender effect was particularly marked when the girls lived in a household without siblings, again in keeping with previous studies (Paul & Serpell, 1992; Westgarth et al., 2013): it may be that siblings spend time with each other, leaving less time to develop attachments to their pets.

The child's development of empathy is also important for their relations with animals in later life. Thus, for example, Meyer, Forkman, and Paul (2014) found that veterinary students with a low level of empathy for animals and with no experience of dogs were likely to assess a dog's aggressive behaviour as more serious than those students with a high level of empathy for animals. Indeed, the evidence strongly suggests that there are cognitive and social advantages for the developing child which may come from contact a with pets. These advantages, in turn, may be to the short- and long-term benefit of the child, those in their family and their social network, and their pets and animals generally. As Jalongo (2012) suggests, more needs to be done to understand the nuances of the attachments that form between children and animals.

The death of a pet

It is not easy to lose someone, say a relative or a friend, to whom a close attachment exists. The same is true for those, children and adults alike, who form a close attachment to their pet: indeed, children may feel the loss of a pet particularly keenly (Schmidt et al., 2018). The death of a pet may precipitate a period of grieving, sometimes for as long as a year (Wrobel & Dye, 2003), as everyday routines are lost while trying to cope with feelings of sadness and even anxiety and depression (Kemp, Jacobs, & Stewart, 2016; Redmalm, 2003). There are similarities in the process of grieving for a human and an animal (Eckerd, Barnett, & Jett-Dias, 2016; Lavorgna & Hutton, 2019) but there are also marked differences. Sharkin and Knox (2003) make the point that the relationship with a companion animal is often unconditional in that the pet offers affection and support in a way seldom found in a human companion. The person who lives alone may feel the loss of their pet particularly keenly. In the normal course of events, the bereaved person will receive some level of social support from relatives and friends. When a person dies the bereaved will normally receive social support from relatives and friends, support which may not be forthcoming when a companion animal dies as not everyone appreciates the nature of the loss (Reisbig, Hafen, Siqueira Drake, Girard, & Breunig, 2017). Indeed, the last thing the owner wants to hear immediately after the death of their pet, no matter how kindly the intention, is "Will you be getting another one?"

Substantial numbers of pets die in road traffic accidents (Wilson, Gruffydd-Jones, & Murray, 2017); when a pet survives an accident the emotional impact may change owner's behaviour so that they become more cautious in safeguarding their pet (Rochlitz, 2004). The manner of the animal's death may also be related to the owner's reaction. Stokes, Planchon, Templer, and Keller (2002) found that when a pet died in an accident, there was a greater likelihood of extended grief compared to when an animal in veterinary care succumbed to an illness.

Hunt and Padilla (2006) developed the Pet Bereavement Questionnaire (PBQ) to provide a means by which professionals and researchers could assess an individual's grief, anger, and guilt at losing a companion animal. PDQ assessments indicate that level of grief correlates strongly with level of attachment to the animal and that the anger and guilt scales correlate with symptoms of depression. Testoni et al. (2019) used a translation of the PBQ alongside similar instruments in looking at standardised assessments for use by vets when assisting owners to make end of life decisions for their pets.

As Christiansen, Kristensen, Lassen, and Sandøe (2015) explain, a veterinarian must manage a complex and difficult situation when faced with an owner and a seriously ill animal. Assuming that owner and vet have the animal's best interest in mind, should the vet provide impartial advice or seek to influence the owner's decision? Should the vet offer their specialist opinion, guided by the law if relevant, or just the prognosis? The animal's quality of life is clearly important in end of life decision making but how is it to be gauged? Should quality of life be judged by the vet, the owner, vet and owner jointly; should a standardised assessment instrument be used (Belshaw, 2018)? Yet further, as more ethical dilemmas present themselves, from a survey of veterinary anaesthetists responsible for critically ill animals, Lehnus, Fordyce, and McMillan (2019) conclude that the majority of respondents are in accord with the British Veterinary Association survey in judging that the veterinary profession should be treating as far as it should rather than as far the limit.

It can be seen that there are overlaps with concerns in the medical treatment of humans such as making end-of-life decisions with respect to quality of life and re-sponsibility for the final decision and, in some countries, the individual's ability to pay for treatment. In humans, there may be informed consent but this is clearly not the case for an animal and payment decisions rest with an may influence the owner (Gray, Fox, & Hobson-West, 2018). It may also fall to the vet to manage the owner's emotions through the process of making decisions about euthanasia and may infleunce and dealing with the ensuing grief (Morris, 2012).

Given the natural constraints on animal consent, is it right to assume that humans can make life and death decisions about animals? Meijer (2018) makes a highly pertinent point in challenging anthropocentric assumptions in the way we make life and death decisions on behalf of our pets: "The word 'euthanasia' is currently used as euphemism for many practices in which other animals are killed, and because this word is used to make their deaths seem beneficial, it functions to obscure, or even legitimate, the vio-lence behind it. Using the right word for these acts, which is often 'killing', is important in challenging this" (p. 217).

It is possible to lose an animal and not know if it remains alive. Some animals wander and become lost and others may be stolen, in both cases the owner is left in a state of limbo wavering between hope and despair. The theft of a dog is a criminal offence dealt with under the same legislation as property theft, although the view of a dog as property is arguable (Harris, 2018). The exact number of stolen dogs is impossible to know as many thefts are unreported. However, information from insurance companies (e.g., www.directline.com/pet-cover/dog-theft) allows a UK annual estimate of almost 2,000 stolen dogs. The most popular targets for thieves include Pit Bull terriers, crossbreeds, Yorkshire terriers, and French bulldogs. Stolen dogs may be resold to a new owner or to a puppy farm for breeding. Finally, companion animals can be lost in natural disasters such as fires and earthquakes. Hunt, Al-Awadi, and Johnson (2008) record how Hurricane Katrina, which hit the gulf coast of the United States in 2005, led to the loss of many animals leaving survivors to cope with the effects of the disaster and to grieve for their pet.

Given the psychological impact of losing a pet it is not surprising that counselling services exist for grieving pet owners. In the UK *The Blue Cross* provide a confidential pet bereavement service and *Cats Protection* offer a confidential telephone line, "Paws to Listen," for grieving cat owners.

The *British Horse Society* offers "Friends at the End" to ensure that horse owners do not have to be alone when losing their horse. It may be that grief becomes protracted and professional help is required. Hess-Holden, Monaghan, and Justice (2017) offer guidelines for mental health professionals to set up a pet bereavement support group.

It would be a mistake to assume that animals are not affected by the death of an animal of the same species. There is evidence that primates such as chimpanzees are aware of death and may even react with empathy when a familiar chimpanzee dies (Pierce, 2013). In other parts of the animal kingdom, there are examples where animals such as birds and aquatic mammals behave in specific ways when a fellow animal is dying. Where there are two or more companion animals, a death may herald a difficult period. Pierce (2013) states that:

> Animals may not outwardly express their grief in ways discernible to us. Sometimes the first response of an animal is acute grief and crying. Some animals show no initial reaction to the death of a companion (human or animal). Later, though, they may begin to search for their loved one, becoming more and more apprehensive and vigilant. Some dogs will show signs of depression, loss of appetite, listlessness. Some will vocalize; others will grow quiet. Some will become clingy; others withdraw. (p. 472)

While our relationships with companion animals may bring pleasure, it is not always plain sailing. There are a variety of problems which may present themselves.

Problems with pets

The type of problem a pet brings will to some degree be dependent upon its age. Thus, young animals need to learn not to soil indoors but to go outside when nature calls, not to scratch and chew the furniture, and to wear a collar and walk on a lead. A failure in early learning will lead to later problems while as they grow older our animals must also cope with visits to the vets, to travel as necessary, learn not to bark at all and sundry, and sometimes to live with other pets. These problems can be a major irritation to owners but there are two types of problems that are major issues for all concerned: these two are aggression and anxiety, each of which will be considered in turn.

References

Amat, M., & Manteca, X. (2019). Common feline problem behaviours: Owner-directed aggression. *Journal of Feline Medicine and Surgery, 21*, 245–255.

Amiot, C., Bastian, B., & Martens, P. (2016). People and companion animals: It takes two to tango. *Bioscience, 66*, 552–560.

Amiot, C. E., & Bastian, B. (2017). Solidarity with animals: Assessing a relevant dimension of social identification with animals. *PLoS ONE, 12*, e0168184.

Anderson, P. K. (2003). A bird in the house: An anthropological perspective on companion parrots. *Society & Animals, 11*, 393–418.

Anderson, P. K. (2014). Social dimensions of the human–avian bond: Parrots and their persons. *Anthrozoös, 27*, 371–387.

Arahori, M., Kuroshima, H., Hori, Y., Takagi, S., Chijiiwa, H., & Fujita, K. (2017). Owners' view of their pets' emotions, intellect, and mutual relationship: Cats and dogs compared. *Behavioural Processes, 141*, 316–321.

Barcelos, A. M., McPeake, K., Affenzeller, N., & Mills, D. S. (2018). Common risk factors for urinary house soiling (periuria) in cats and its differentiation: The sensitivity and specificity of common diagnostic signs. Frontiers in Veterinary Science, *5*, 108.

Beck, A. M. (2014). The biology of the human–animal bond. *Animal Frontiers, 4*, 32–36.

Belshaw, Z. (2018). Quality of life assessment in companion animals: What, why, who, when and how. *Companion Animal, 23*, 264–268.

Bennett, P. C., Rutter, N. J., Woodhead, J. K., & Howell, T. J. (2017). Assessment of domestic cat personality, as perceived by 416 owners, suggests six dimensions. *Behavioural Processes, 141*, 273–283.

Bensky, M. K., Gosling, S. D., & Sinn, D. L. (2013). The world from a dog's point of view: A review and synthesis of dog cognition research. *Advances in the Study of Behavior, 45*, 209–406.

Bernstein, P. L. (2007). The human-cat relationship. In I. Rochlitz (Ed.), *The welfare of cats* (pp. 47–89). Dordrecht, The Netherlands: Springer.

Blouin, D. D. (2013). Are dogs children, companions, or just animals? Understanding variations in people's orientations toward animals. *Anthrozoös, 26*, 279–294.

Boruta, A., Kurek, A., & Lewandowska, M. (2016). The criteria for choosing a companion dog. *Annals of Warsaw University of Life Sciences-SGGW, Animal Science, 55*, 147–156.

Bowlby, J. (1953). *Child care and the growth of love*. Harmondsworth, Middlesex: Penguin Books.

Bowlby, J. (1956). Mother-child separation. *Mental Health and Infant Development, 1*, 117–122.

Bradshaw, J. W. S. (2016). Sociality in cats: A comparative review. *Journal of Veterinary Behavior, 11*, 113–124.

Brown, S. G., & Rhodes, R. E. (2006). Relationships among dog ownership and leisure-time walking in Western Canadian adults. *American Journal of Preventive Medicine, 30*, 131–136.

Brown, W. P., Davidson, J. P., & Zuefle, M. E. (2013). Effects of phenotypic characteristics on the length of stay of dogs at two no kill animal shelters. *Journal of Applied Animal Welfare Science, 16*, 2–18.

Cannas, S., Rampini, F., Levi, D., & Costa, E. D. (2014). Shelter dogs and their destiny. A retrospective analysis to identify predictive factors: A pilot study. *Macedonian Veterinary Review, 37*, 151–156.

Carter, S. B. (2016). Establishing a framework to understand the regulation and control of dogs in urban environments: A case study of Melbourne, Australia. *SpringerPlus, 5*, 1190.

Charles, N. (2014). Animals just love you as you are: Experiencing kinship across the species barrier. *Sociology, 48*, 715–730.

Christian, H., Trapp, G., Lauritsen, C., Wright, K., & Giles-Corti, B. (2013). Understanding the relationship between dog ownership and children's physical activity and sedentary behaviour. *Pediatric Obesity, 8*, 392–403.

Christian, H., Bauman, A., Epping, J. N., Levine, G. N., McCormack, G., Rhodes, R. E.,…. Westgarth, C. (2016). Encouraging dog walking for health promotion and disease prevention. *American Journal of Lifestyle Medicine, 12*, 233–243.

Christiansen, S. B., Kristensen, A. T., Lassen, J., & Sandøe, P. (2015). Veterinarians' role in clients' decision-making regarding seriously ill companion animal patients. *Acta Veterinaria Scandinavica*, *58*, 1–14.

Chung, T., Park, C., Kwon, Y., & Yeon, S. (2016). Prevalence of canine behavior problems related to dog-human relationship in South Korea – A pilot study. *Journal of Veterinary Behavior*, *11*, 26–30.

Coleman, K. J., Rosenberg, D. E., Conway, T. L., Sallis, J. F., Saelens, B. E., Frank, L. D., & Cain, K. (2008). Physical activity, weight status, and neighborhood characteristics of dog walkers. *Preventive Medicine*, *47*, 309–312.

Cozzi, A., Mariti, C., Ogi, A., Sighieri, C., & Gazzano, A. (2016). Behavioral modification in sheltered dogs. *Dog Behavior*, *2*, 1–12.

Crawford, E. K., Worsham, N. L., & Swinehart, E. R. (2006). Benefits derived from companion animals, and the use of the term "attachment." *Anthrozoös*, *19*, 98–112.

Cutt, H., Giles-Corti, B., Knuiman, M., Timperio, A., & Bull, F. (2008). Understanding dog owners' increased levels of physical activity: Results From RESIDE. *American Journal of Public Health*, *98*, 66–69.

Dalla Villa, P., Iannetti, L., Podaliri Vulpiani, M., Maitino, A., Trentini, R., & Del Papa, S. (2008). A management model applied in two no-kill dog shelters in central Italy: Use of population medicine for three consecutive years. *Veterinaria Italiana*, *44*, 347–359.

Demirbas, Y. S., Safak, E., Emre, B., Piskin, İ., Ozturk, H., & Pereira, G. D. G. (2017). Rehabilitation program for urban free–ranging dogs in a shelter environment can improve behavior and welfare. *Journal of Veterinary Behavior: Clinical Applications and Research*, *18*, 1–6.

DePorter, T. L., & Elzerman, A. L. (2019). Common feline problem behaviors: Destructive scratching. *Journal of Feline Medicine and Surgery*, *21*, 235–243.

Diesel, G., Brodbelt, D., & Pfeiffer, D. U. (2010). Characteristics of relinquished dogs and their owners at 14 rehoming centers in the United Kingdom. *Journal of Applied Animal Welfare Science*, *13*, 15–30.

Duffy, D. L., de Moura, R. T. D., & Serpell, J. A. (2017). Development and evaluation of the Fe-BARQ: A new survey instrument for measuring behavior in domestic cats (*Felis s. catus*). *Behavioural Processes*, *141*, 329–341.

Duranton, C., Bedossa, T., & Gaunet, F. (2018). Pet dogs synchronize their walking pace with that of their owners in open outdoor areas. *Animal Cognition*, *21*, 219–226.

Echeverria, A., Karp, D. S., Naidoo, R., Zhao, J., & Chan, K. (2018). Approaching human-animal relationships from multiple angles: A synthetic perspective. *Biological Conservation*, *224*, 50–62.

Eckerd, L. M., Barnett, J. E., & Jett-Dias, L. (2016). Grief following pet and human loss: Closeness is key. *Death Studies*, *40*, 275–282.

Edensor, T. (2000). Walking in the British countryside: Reflexivity, embodied practices and ways to escape. *Body & Society*, *6*, 81–106.

Eriksson, P., Loberg, J., & Andersson, M. (2009). A survey of cat shelters in Sweden. *Animal Welfare*, *18*, 283–288.

Evans, R., Lyons, M., Brewer, G., & Tucci, S. (2019). The purrfect match: The influence of personality on owner satisfaction with their domestic cat (*Felis silvestris catus*). *Personality and Individual Differences*, *138*, 252–256.

Evans-Wilday, A. S., Hall, S. S., Hogue, T. E., & Mills, D. S. (2018). Self-disclosure with dogs: Dog owners' and non-dog owners' willingness to disclose emotional topics, *Anthrozoös*, *31*, 353–366.

Fantuzzi, J. M., Miller, K. A., & Weiss, E. (2010). Factors relevant to adoption of cats in an animal shelter. *Journal of Applied Animal Welfare Science*, *13*, 174–179.

Finka, L. R., Ward, J., Farnworth, M. J., & Mills, D. S. (2019). Owner personality and the wellbeing of their cats share parallels with the parent-child relationship. *PLoS ONE*, *14*, e0211862.

Fletcher, T., & Platt, L. (2018). (Just) a walk with the dog? Animal geographies and negotiating walking spaces. *Social & Cultural Geography*, *19*, 211–229.

Foyer, P., Bjällerhag, N., Wilsson, E., & Jensen, P. (2014). Behaviour and experiences of dogs during the first year of life predict the outcome in a later temperament test. *Applied Animal Behaviour Science*, *155*, 93–100.

Gadomski, A. M., Scribani, M. B., Krupa, N., & Jenkins, P. (2017). Pet dogs and child physical activity: The role of child-dog attachment. *Pediatric Obesity, 12,* e37–e40.

Garrison, L., & Weiss, E. (2015). What do people want? Factors people consider when acquiring dogs, the complexity of the choices they make, and implications for nonhuman animal relocation programs. *Journal of Applied Animal Welfare Science, 18,* 57–73.

Gates, M. C., Zito, S., Thomas, J., & Dale, A. (2018). Post-adoption problem behaviours in adolescent and adult dogs rehomed through a New Zealand animal shelter. *Animals, 8,* 93.

Ghirlanda, S., Acerbi, A., Herzog, H., & Serpell, J. A. (2013). Fashion vs. function in cultural evolution: The case of dog breed popularity. *PLoS ONE, 8,* e74770.

González-Martínez, A., Martínez, M. F., Rosado, B., Luño, I., Santamarina, G., Suárez, M. L.,… Diéguez, F. J. (2019). Association between puppy classes and adulthood behavior of the dog. *Journal of Veterinary Behavior, 32,* 36–41.

Gosling, S. D., Sandy, C. J., & Potter, J. (2010). Personalities of self-identified "dog people" and "cat people." *Anthrozoös, 23,* 213–222.

Graham, T. M., & Glover, T. D. (2014). On the fence: Dog parks in the (un)leashing of community and social capital. *Leisure Sciences, 36,* 217–234.

Gray, C., Fox, M., & Hobson-West, P. (2018). Reconciling autonomy and beneficence in treatment decision-making for companion animal patients. *Liverpool Law Review, 39,* 47–69.

Gray, R., Butler, S., Douglas, C., & Serpell, J. (2016). Puppies from "puppy farms" show more temperament and behavioural problems than if acquired from other sources. Paper presented at *Recent Advances in Animal Welfare Science V: UFAW Animal Welfare Conference.* Newcastle University, Newcastle upon Tyne.

Greenebaum, J. (2004). It's a dog's life: Elevating status from pet to "fur baby" at yappy hour. *Society & Animals, 12,* 117–135.

Guagnin, M., Perri, A. R., & Petraglia, M. D. (2018). Pre-Neolithic evidence for dog-assisted hunting strategies in Arabia. *Journal of Anthropological Archaeology, 49,* 225–236.

Gunter, L. M., Barber, R. T., & Wynne, C. D. L. (2016). What's in a name? Effect of breed perceptions & labeling on attractiveness, adoptions & length of stay for pit-bull-type dogs. *PLoS ONE, 11,* e0146857.

Harris, L. (2018). Dog theft: A case for tougher sentencing legislation. *Animals, 8,* 78.

Hawes, S., Kerrigan, J., & Morris, K. (2018). Factors informing outcomes for older cats and dogs in animal shelters. *Animals, 8,* 36.

Hawkins, R. D., Williams, J. M., & Scottish Society for the Prevention of Cruelty to Animals (Scottish SPCA). (2017). Childhood attachment to pets: Associations between pet attachment, attitudes to animals, compassion, and humane behaviour. *International Journal of Environmental Research in Public Health, 14,* 490–510.

Heath, S. (2019). Common feline problem behaviours: Unacceptable indoor elimination. *Journal of Feline Medicine and Surgery, 21,* 199–208.

Herzog, H. (2007). Gender differences in human-animal interactions: A review. *Anthrozoös, 20,* 7–21.

Herzog, H. (2014). Biology, culture, and the origins of pet-keeping. *Animal Behavior and Cognition, 1,* 296–308.

Hess-Holden, C. L., Monaghan, C. L., & Justice, C. A. (2017). Pet bereavement support groups: A guide for mental health professionals. *Journal of Creativity in Mental Health, 12,* 440–450.

Hirschenhauser, K., Meichel, Y., Schmalzer, S., & Beetz, A. M. (2017). Children love their pets: Do relationships between children and pets co-vary with taxonomic order, gender, and age? *Anthrozoös, 30,* 441–456.

Horwitz, D. F. (2019). Common feline problem behaviors: Urine spraying. *Journal of Feline Medicine and Surgery, 21,* 209–219.

Hosey, G., & Melfi, V. (2014). Human-animal interactions, relationships and bonds: A review and analysis of the literature. *International Journal of Comparative Psychology, 27,* 117–142.

Howell, T. J., Bowen, J., Fatjó, J., Calvo, P., Holloway, A., & Bennett, P. C. (2017). Development of the cat-owner relationship scale (CORS). *Behavioural Processes, 141,* 305–315.

Hsu, Y., Severinghaus, L. L., & Serpell, J. A. (2003). Dog keeping in Taiwan: Its contribution to the problem of free-roaming dogs. *Journal of Applied Animal Welfare Science*, 6, 1–23.

Hunt, M., Al-Awadi, H., & Johnson, M. (2008). Psychological sequelae of pet loss following Hurricane Katrina. *Anthrozoös*, 21, 109–121.

Hunt, M., & Padilla, Y. (2006). Development of the Pet Bereavement Questionnaire. *Anthrozoös*, 19, 308–324.

Instone, L., & Mee, K. (2011). Companion acts and companion species: Boundary transgressions and the place of dogs in urban public space. In J. Bull (Ed.), *Animal movements • Moving animals. Essays on direction, velocity and agency in humanimal encounters* (pp. 229–250). Centre for Gender Research, Uppsala University, Uppsala, Sweden.

Instone, L., & Sweeny, J. (2014). *Placing companion animals in the city: Towards the constructive cohabitation of humans and dogs in urban areas*. Project Report, Centre for Urban and Regional Studies (CURS), The University of Newcastle, Callaghan, New South Wales, Australia.

Irvine, L. (2009). *Filling the ark: Animal welfare in disasters*. Philadelphia, PA: Temple University Press.

Jalongo, M. (2012). An attachment perspective on the child-dog bond: Interdisciplinary and international research findings. *Early Childhood Education Journal*, 43, 395–405.

Jalongo, M. R. (Ed.). (2018). *Children, dogs and education: Caring for, learning alongside, and gaining support from canine companions*. Cham, Switzerland: Springer.

John, O. P., Naumann, L., & Soto, C. J. (2008). The Big Five trait taxonomy: Discovery, measurement, and theoretical issues. In O. P. John, R. W. Robins, and L. A. Pervin (Eds.), *Handbook of personality: Theory and research* (3rd ed., pp. 114–158). New York: Guilford Press.

Kellert, S. R., & Berry, J. K. (1987). Attitudes, knowledge, and behaviors toward wildlife as affected by gender. *Wildlife Society Bulletin*, 15, 363–371.

Kemp, H. R., Jacobs, N., & Stewart, S. (2016). The lived experience of companion-animal loss: A systematic review of qualitative studies. *Anthrozoös*, 29, 533–557.

Kerr, C., Rand, J., Morton, J., Reid, R., & Paterson, M. (2018). Changes associated with improved outcomes for cats entering RSPCA Queensland Shelters from 2011 to 2016. *Animals*, 8, 95.

Kirk, C. P. (2019). Dogs have masters, cats have staff: Consumers' psychological ownership and their economic valuation of pets. *Journal of Business Research*, 99, 306–318.

Kotrschal, K. (2018). How wolves turned into dogs and how dogs are valuable in meeting human social needs. *People and Animals: The International Journal of Research and Practice*, 1, Article 6.

Lakestani, N. N., Donaldson, M. L., & Waran, N. (2014). Interpretation of dog behavior by children and young adults. *Anthrozoös*, 27, 65–80.

Lavorgna, B. F., & Hutton, V. E. (2019). Grief severity: A comparison between human and companion animal death. *Death Studies*, 43, 521–526.

Lehnus, K. S., Fordyce, P. S., & McMillan, M. W. (2019). Ethical dilemmas in clinical practice: A perspective on the results of an electronic survey of veterinary anaesthetists. *Veterinary Anaesthesia and Analgesia*, 46, 260–275.

Litchfield, C. A., Quinton, G., Tindle, H., Chiera, B., Kikillus, K. H., & Roetman, P. (2017). The 'Feline Five': An exploration of personality in pet cats (*Felis catus*). *PLoS ONE*, 12, e0183455.

Maharaj, N., & Haney, C. J. (2015). A qualitative investigation of the significance of companion dogs. *Western Journal of Nursing Research*, 37, 1175–1193.

Mariti, C., Carlone, B., Ricci, E., Sighieri, C., & Gazzano, A. (2014). Intraspecific attachment in adult domestic dogs (*Canis familiaris*): Preliminary results. *Applied Animal Behaviour Science*, 152, 64–72.

Marsa-Sambola, F., Muldoon, J., Williams, J., Lawrence, A., Connor, M., & Currie, C. (2016b). The short attachment to pets scale (SAPS) for children and young people: Development, psychometric qualities and demographic and health associations. *Child Indicators Research*, 9, 111–131.

Marsa-Sambola, F., Williams, J., Muldoon, J., Lawrence, A. B., Connor, M., Roberts, C..... Currie, C. (2016a). Sociodemographics of pet ownership among adolescents in Great Britain: Findings from the HBSC study in England, Scotland and Wales. *Anthrozoos*, 29, 559–580.

McMillan, F. D. (2017). Behavioral and psychological outcomes for dogs sold as puppies through pet stores and/or born in commercial breeding establishments: Current knowledge and putative causes. *Journal of Veterinary Behavior, 19*, 14–26.

McMillan, F. D., Duffy, D. L., & Serpell, J. A. (2011). Mental health of dogs formerly used as "breeding stock" in commercial breeding establishments. *Applied Animal Behaviour Science, 135*, 86–94.

McMillan, F. D., Serpell, J. A., Duffy, D. L., Masaoud, E., & Dohoo, I. R. (2013). Differences in behavioral characteristics between dogs obtained as puppies from pet stores and those obtained from noncommercial breeders. *Journal of the American Veterinary Medical Association, 242*, 1359–1363.

McNicholas, J., & Collis, G. M. (2000). Dogs as catalysts for social interactions: Robustness of the effect. *British Journal of Psychology, 91*, 61–70.

Meijer, E. (2018). The good life, the good death: Companion animals and euthanasia. *Animal Studies Journal, 7*, 205–225.

Meyer, I., Forkman, B., & Paul, E. (2014). Factors affecting the human interpretation of dog behavior. *Anthrozoös, 27*, 127–140.

Morey, D. F. (2006). Burying key evidence: The social bond between dogs and people. *Journal of Archaeological Science, 33*, 158–175.

Morris, P. (2012). Managing pet owners' guilt and grief in veterinary euthanasia encounters. *Journal of Contemporary Ethnography, 41*, 337–365.

Muldoon, J. C., Williams, J. M., & Lawrence, A. (2016). Exploring children's perspectives on the welfare needs of pet animals. *Anthrozoös, 29*, 357–375.

Murray, J. K., Browne, W. J., Roberts, M. A., Whitmarsh, A., & Gruffydd-Jones, T. J. (2010). Number and ownership profiles of cats and dogs in the UK. *The Veterinary Record, 166*, 163–168.

Murugan, M. M. S., Arunvikram, K., Pavulraj, S., Milton, A. A. P., Sinha, D. K., & Singh, B. R. (2015). Companion animals: A potential threat in emergence and transmission of parasitic zoonoses. *Advances in Animal Veterinary Sciences, 3*, 594–604.

Nast, H. J. (2006). Loving…. whatever: Alienation, neoliberalism and pet-love in the twenty-first century. *ACME: An International E-Journal for Critical Geographies, 5*, 300–327.

Newell, B. R., Lagnado, D. A., & Shanks, D. R. (2015). *Straight choices: The psychology of decision making* (2nd ed.). New York: Psychology Press.

Normando, S., Stefanini, C., Meers, L., Adamelli, S., Coultis, D., & Bono, G. (2006). Some factors influencing adoption of sheltered dogs. *Anthrozoös, 19*, 211–224.

Odendaal, J. S. J., & Meintjes, R.A. (2003). Neurophysiological correlates of affiliative behaviour between humans and dogs. *Veterinary Journal, 165*, 296–301.

Palacio, J., León-Artozqui, M., Pastor-Villalba, E., Carrera-Martín, F., & García-Belenguer, S. (2007). Incidence of and risk factors for cat bites: A first step in prevention and treatment of feline aggression. *Journal of Feline Medicine and Surgery, 9*, 188–195.

Patronek, G. L., & Crowe, A. (2018). Factors associated with high live release for dogs at a large, open-admission, municipal shelter. *Animals, 8*, 45.

Paul, E. S., & Serpell, J. A. (1966). Obtaining a new pet dog: Effects on middle childhood children and their families. *Applied Animal Behaviour Science, 47*, 17–29.

Paul, E. S., & Serpell, J. A. (1992). Why children keep pets: The influence of child and family characteristics. *Anthrozoös, 5*, 231–244.

Paul, E. S., & Serpell, J. A. (1993). Childhood pet keeping and humane attitudes in young adulthood. *Animal Welfare, 2*, 321–337.

Penar, W., & Klocek, C. (2018). Aggressive behaviors in domestic cats (*Felis catus*). *Annals of Warsaw University of Life Sciences – SGGW, Animal Science, 57*, 143–150.

Phillips, C., Izmirli, S., Aldavood, J., Alonso, M., Choe, B., Hanlon, A.… Rehn, T. (2011). An international comparison of female and male students' attitudes to the use of animals. *Animals, 1*, 7–26.

Pierce, J. (2013). The dying animal. *Journal of Bioethical Inquiry, 10*, 469–478.

Podberscek, A. L. (2009). Good to pet and eat: The keeping and consuming of dogs and cats in South Korea. *Journal of Social Issues, 65*, 615–632.

Pongrácz, P., & Szapu, J. S. (2018). The socio-cognitive relationship between cats and humans – Companion cats (*Felis catus*) as their owners see them. *Applied Animal Behaviour Science, 207*, 57–66.
Powell, L., Chia, D., McGreevy, P., Podberscek, A. L., Edwards, K. M., Neilly, B....... Stamatakis, E. (2018). Expectations for dog ownership: Perceived physical, mental and psychosocial health consequences among prospective adopters. *PloS ONE, 13*, e0200276.
Protopopova, A., Gilmour, A. J., Weiss, R. H., Shen, J. Y., & Wynne, C. D. L. (2012). The effects of social training and other factors on adoption success of shelter dogs. *Applied Animal Behaviour Science, 142*, 61–68.
Protopopova, A., Hauser, H., Goldman, K. J., & Wynne, C. D. L. (2018). The effects of exercise and calm interactions on in-kennel behavior of shelter dogs. *Behavioural Processes, 146*, 54–60.
Protopopova, A., Mehrkam, L. R., Boggess, M. M., & Wynne, C. D. L. (2014). In-kennel behavior predicts length of stay in shelter dogs. *PloS ONE, 9*, e114319.
Protopopova, A., & Wynne, C. D. L. (2015). Improving in-kennel presentation of shelter dogs through response dependent and independent treat delivery. *Journal of Applied Behavior Analysis, 48*, 1–12.
Protopopova, A., & Wynne, C. D. L. (2016). Judging a dog by its cover: Morphology but not training influences visitor behavior toward kenneled dogs at animal shelters. *Anthrozoös, 29*, 469–487.
Purewal, R., Christley, R., Kordas, K., Joinson, C., Meints, K., Gee, N., & Westgarth, C. (2017). Companion animals and child/adolescent development: A systematic review of the evidence. *International Journal of Environmental Research and Public Health, 14*, 234.
Rahim, T., Barrios, P. R., McKee, G., McLaws, M., & Kosatsky, T. (2018). Public health considerations associated with the location and operation of off-leash dog parks. *Journal of Community Health, 43*, 433–440.
Rand, J., Lancaster, E., Inwood, G., Cluderay, C., & Marston, L. (2018). Strategies to reduce the euthanasia of impounded dogs and cats used by councils in Victoria, Australia. *Animals, 8*, 100–135.
Redmalm, D. (2003). Pet grief: When is non-human life grievable? *Sociological Review, 63*, 19–35.
Redmalm, D. (2014). Holy bonsai wolves: Chihuahuas and the Paris Hilton syndrome. *International Journal of Cultural Studies, 17*, 93–109.
Reese, L. A., Skidmore, M., Dyar, W., & Rosebrook, E. (2017). No dog left behind: A hedonic pricing model for animal shelters. *Journal of Applied Animal Welfare Science, 20*, 52–64.
Rehn, T., Lindholm, U., Keeling, L., & Forkman, B. (2014). I like my dog, does my dog like me? *Applied Animal Behaviour Science, 150*, 65–73.
Reisbig, A. M., Hafen Jr, M., Siqueira Drake, A. A., Girard, D., & Breunig, Z. B. (2017). Companion animal death: A qualitative analysis of relationship quality, loss, and coping. *OMEGA-Journal of Death and Dying, 75*, 124–150.
Řezáč, P., Viziová, P., Dobešová, M., Havlíček, Z., & Pospíšilová, D. (2011). Factors affecting dog–dog interactions on walks with their owners. *Applied Animal Behaviour Science, 134*, 170–176.
Rioja-Lang, F., Bacon, H., Connor, M., & Dwyer, C. M. (2019). Determining priority welfare issues for cats in the United Kingdom using expert consensus. *Veterinary Record Open, 6*, e000365.
Roberts, J. D., Rodkey, L., Grisham, C., & Ray, R. (2017). The influence of family dog ownership and parental perceived built environment measures on children's physical activity within the Washington, DC area. *International Journal of Environmental Research and Public Health, 14*, 1398–1409.
Rochlitz, I. (2004). The effects of road traffic accidents on cats and their owners. *Animal Welfare, 13*, 51–55.
Rock, M. J., Degeling, C., Graham, T. M., Toohey, A. M., Rault, D., & McCormack, G. R. (2016). Public engagement and community participation in governing urban parks: A case study in changing and implementing a policy addressing off-leash dogs. *Journal of Critical Public Health, 26*, 588–601.
Rook, D. (2014). Who gets Charlie? The emergence of pet custody disputes in family law: Adapting theoretical tools from child law. *International Journal of Law, Policy and the Family, 28*, 177–193.
Rook, D. (2018). For the love of Darcie: Recognising the human–companion animal relationship in housing law and policy. *Liverpool Law Review, 39*, 29–46.
Sandøe, P., Kondrup, S. V., Bennett P. C., Forkman, B., Meyer I, Proschowsky, H. F..... Lund, T. B. (2017). Why do people buy dogs with potential welfare problems related to extreme conformation

and inherited disease? A representative study of Danish owners of four small dog breeds. *PLoS ONE*, *12*, e0172091.

Schmidt, M., Naylor, P. E., Cohen, D., Gomez, R., Moses Jr., J. A., Rappoport, M., & Packman, W. (2018). Pet loss and continuing bonds in children and adolescents. *Death Studies*, 44, 278–284.

Schumaker, B. A., Miller, M. M., Grosdidier, P., Cavender, J. L., Montgomery, D. L., Cornish, T. E., … O'Toole, D. (2012). Canine distemper outbreak in pet store puppies linked to a high-volume dog breeder. *Journal of Veterinary Diagnostic Investigation*, *24*, 1094–1098.

Sediva, M., Holcova, K., Pillerova, L., Koru, E., & Řezáč, P. (2017). Factors influencing off-leash dog walking in public places. *Acta Universitatis Agriculturae et Silviculturae Mendelianae Brunensis*, *65*, 1761–1766.

Sevillano, V., & Fiske, S. T. (2016). Animals as social objects: Groups, stereotypes, and intergroup threats. *European Psychologist*, *21*, 206–217.

Sharkin, B., & Knox, D. (2003). Pet loss: Issues and implications for the psychologist. *Professional Psychology: Research and Practice*, 34, 414–421.

Shore, E. R. (2005). Returning a recently adopted companion animal: Adopters' reasons for and reactions to the failed adoption experience. *Journal of Applied Animal Welfare Science*, *8*, 187–198.

Statista. (2018). Leading pets, ranked by household ownership in the United Kingdom (UK) in 2017/18. www.statista.com/statistics/308218/leading-ten-pets-ranked-by-household-ownership-in-the-united-kingdom-uk.

Statts, S., Sears, K., & Pierfelice, L. (2006). Teachers' pets and why they have them: An investigation of the human animal bond. *Journal of Applied Social Psychology*, *36*, 1881–1891.

Stella, J. L., & Croney, C. C. (2016). Environmental aspects of domestic cat care and management: Implications for cat welfare. *Scientific World Journal*, *2016*, 1–7.

Sterneberg-Van der Maaten, T., Turner, D., Van Tilburg, J., & Vaarten, J. (2016). Benefits and risks for people and livestock of keeping companion animals: Searching for a healthy balance. *Journal of Comparative Pathology*, *155*, S8–S17.

Stokes, S., Planchon, L., Templer, D., & Keller, J. (2002). Death of a companion cat or dog and human bereavement: Psychosocial variables. *Society & Animals*, *10*, 93–105.

Testoni, I., De Cataldo, L., Ronconi, L., Colombo, E. S., Stefanini, C., Dal Zotto, B., & Zamperini, A. (2019). Pet grief: Tools to assess owners' bereavement and veterinary communication skills. *Animals*, *9*, 67.

Thompson, K. L., & Gullone, E. (2008). Prosocial and antisocial behaviors in adolescents: An investigation into associations with attachment and empathy. *Anthrozoös*, *21*, 123–137.

Thomson, J. E., Hall, S. S., & Mills, D. S. (2018). Evaluation of the relationship between cats and dogs living in the same home. *Journal of Veterinary Behavior*, *27*, 35–40

Tipper, B. (2011a). Pets and personal life. In V. May (Ed.), *Sociology of personal life* (pp. 85–97). Basingstoke, Hants: Palgrave.

Tipper, B. (2011b). "A dog who I know quite well": Everyday relationships between children and animals. *Children's Geographies*, *9*, 145–165.

Tipper, B. (2013). Moments of being and ordinary human-animal encounters. *Virginia Woolf Miscellany*, *84*, 14–16.

Turner, D. C. (2017). A review of over three decades of research on cat-human and human-cat interactions and relationships. *Behavioural Processes*, *141*, 297–304.

Udell, M. A. R., & Wynne, C. D. L. (2008). A review of domestic dogs' (*Canis Familiaris*) human-like behaviours: Or why behavior analysts should stop worrying and love their dogs. *Journal of the Experimental Analysis of Behavior*, *89*, 247–261.

Vänskä, A. (2016). "Cause I wuv you!" Pet dog fashion and emotional consumption. *Ephemera: Theory & Politics in Organization*, *16*, 75–97.

Vigne, J-D., Guilaine, J., Debue, K., Haye, L., & Gérard, P. (2004). Early taming of the cat in Cyprus. *Science*, *304*(5668), 259–259.

Vilà, C., Savolainen, P., Maldonado, J. E., Amorim, I. R., Rice, J. E., Honeycutt, R. L....., Wayneet, R. K. (1997). Multiple and ancient origins of the domestic dog. *Science*, *276*, 1687–1689.

Vitulová, S., Voslářová, E., Večerek, V., & Bedáňová, I. (2018). Behaviour of dogs adopted from an animal shelter. *Acta Veterinaria Brno*, *87*, 155–163.

Wang, G. D., Zhai, W., Yang, H. C., Wang, L., Zhong, L., Liu, Y. H.,… & Irwin, D. M. (2016). Out of southern East Asia: The natural history of domestic dogs across the world. *Cell Research*, *26*, 21–33.

Wells, D. L., & Hepper, P. G. (2001). The behavior of visitors towards dogs housed in an animal rescue shelter. *Anthrozoös*, *14*, 12–18.

Westgarth, C., Boddy, L. M., Stratton, G., German, A. J., Gaskell, R. M., Coyne, K. P.,… Dawson, S. (2013). Pet ownership, dog types and attachment to pets in 9–10-year-old children in Liverpool, UK. *BMC Veterinary Research*, *9*, 102.

Westgarth, C., Christley, R. M., Jewell, C., German, A. J., Boddy, L. M., & Christian, H. E. (2019). Dog owners are more likely to meet physical activity guidelines than people without a dog: An investigation of the association between dog ownership and physical activity levels in a UK community. *Scientific Reports*, *9*, Article Number 5704.

Westgarth, C., Christley, R., Marvin, G., & Perkins, E. (2017). I walk my dog because it makes me happy: A qualitative study to understand why dogs motivate walking and improved health. *International Journal of Environmental Research and Public Health*, *14*, 936.

Weston, M. A., Fitzsimons, J. A., Wescott, G., Miller, K. K., Ekanayake, K. B., & Schneider, T. (2014). Bark in the park: A review of domestic dogs in parks. *Environmental Management*, *54*, 373–382.

Wilson, J. L., Gruffydd-Jones, T. J., & Murray, J. K. (2017). Risk factors for road traffic accidents in cats up to age 12 months that were registered between 2010 and 2013 with the UK pet cat cohort ('Bristol Cats'). The *Veterinary Record*, *180*, 195.

Wright, J. C., Smith, A., Daniel, K., & Adkins, K. (2007). Dog breed stereotype and exposure to negative behavior: Effects on perceptions of adoptability. *Journal of Applied Animal Welfare Science*, *10*, 255–265.

Wrobel, T. A., & Dye, A. L. (2003). Grieving pet death: Normative, gender, and attachment issues. *OMEGA-Journal of Death and Dying*, *47*, 385–393.

Yong, M. H., & Ruffman, T. (2016). Domestic dogs and human infants look more at happy and angry faces than sad faces. *Multisensory Research*, *29*, 749–771.

Žák, J., Voslářová, E., Večerek, V., & Bedáňová, I. (2015). Sex, age and size as factors affecting the length of stay of dogs in Czech shelters. *Acta Veterinaria Brno*, *84*, 407–413.

Zasloff, R. L. (1996). Measuring attachment to companion animals: A dog is not a cat is not a bird. *Applied Animal Behaviour Science*, *47*, 43–48.

3 Pet problems

Aggression

There are many ways in which animals bring harm to people. van Delft, Thomassen, Schreuder, and Sosef (2019) reviewed animal-related admissions to an emergency department in The Netherlands over a year. There were 516 patients, mainly female, who received treatment because of animal-related injuries, most often fractures and contusions from falling off or tripping over an animal. The animals most often involved were horses, dogs, and cats; three animals, a horse, a rabbit, and a dog, died in the accident. Yet further, those working on farms are at risk of cow-related traumas brought about by kicking, head-butts, and trampling (Murphy, McGuire, O'Malley, & Harrington, 2010). However, pet with a problem is a pet in potential trouble, as Cannas et al. (2018) state: "One of the main reasons dogs are given away, abandoned, or euthanized is a behavioral problem" (a pet with, p. 43). Of the various behavioural problems displayed by pets it is aggression and biting which bring the most consternation and there is much to be gained, by animals and people, in understanding and working on pet problems. While the main concern is with dogs, cats can also be aggressive both towards other cats and to their owners (Amat & Manteca, 2019; Ramos, 2019).

An Australian study by Col, Day, and Phillips (2016) looked at the reasons why dogs were referred to a behaviour problems clinic. From an analysis of 7,858 dogs that showed 11,521 behaviour problems, Col, Day, and Phillips distilled the various problematic behaviours into 22 different classes. A selection of these classes is shown in Table 3.1.

The most frequent issues in the Col, Day, and Phillips study can be set into context by considering what is known about the dog's characteristics – itself a complex amalgam of the various influences in a breed's history (Svartberg, 2006) – and what may be expected in terms of their relative everyday behaviour (Asp, Fikse, Nilsson, & Strandberg, 2015; Takeuchi & Mori, 2006, 2009; Tonoike et al., 2015). In general terms, it may be anticipated, for example, that a Labrador would be more playful than a German Shepherd which, in turn, would be less demanding of its owner than a Cavalier King Charles Spaniel. In addition, as compared to bitches, dogs are more aggressive to other dogs and more liable to snap at children, while bitches are easier to train.

Takeuchi and Mori (2006) provide a comparison of the behavioural traits of pedigree dogs as recorded in Japan, the UK, and the USA. Of course, as Wilsson (2016) makes clear, an animal's given pedigree, the *genotype*, interacts with the environment to give, the *phenotype*, the behaviour that is actually observed. The environment includes the myriad of influences from each individual animal's experience of raising and training. It follows that the vagaries of environmental influences, including the owner's behaviour, act to produce behavioural variations within breeds (Lofgren et al., 2014; Mehrkam & Wynne, 2014).

Table 3.1 Most and least frequent types of dog behaviour problems (from Col et al. 2016)

Problem	No. of dogs (N = 7,858)
Aggression	2,898
Barking	1,101
Anxiety	919
Fear of noises	388
House soiling	377
Digging	29
Chewing	28
Coprophagia (eating faeces)	27
Pica (eating non-food items)	24
Howling	19

The importance of the owner's behaviour in understanding the animal's problematic behaviour – accepting that not all owners see problems in the same way (Pirrone, Pierantoni, Mazzola, Vigo, & Albertini, 2015) – is highlighted by a study of the problems presented by a sample of 737 dogs (Jagoe & Serpell, 1996). Those owners who commit time and expense to engage in obedience training with their dogs experienced fewer problems including both aggression and separation anxiety. On the other hand, those owners who allowed their dog to sleep in their bedroom experienced problems with both aggression and the dog's separation anxiety.

Cannas et al. (2018) classified the problems exhibited by dogs referred to a behaviour clinic under the broad headings of *aggressive* or *anxious*. The anxious dogs were small or medium in size, likely to be young females, to have been obtained from a pet shop, started to show problems within a week of ownership, and were more likely to sleep on the owner's bed. The aggressive dogs were mostly male, typically neutered, obtained from another person such as a dog breeder, and began to show aggressive behaviour after about 4 months. A display of mounting behaviours directed towards people was evident in about one-quarter of the dogs; of these particular dogs, about two-thirds were of the anxious type and about one-third of the aggressive type. The owners of the anxious dogs were significantly more likely to take the dog to a shelter than were those with an aggressive dog.

While dogs do display other problematic behaviours, such as excessive barking, jumping up at people, and wandering off on their own, the two main concerns are with aggression, including biting, and anxiety. This chapter considers aggression and the next looks at anxiety.

Canine aggression

All dogs have the capacity to be aggressive to a greater or lesser extent but certain breeds and some individual dogs are more likely than others to display aggressive behaviour. The consequences of canine aggression for the victim can range from a mild shock to the extreme of trauma and phobia, from a scratch to serious physical injury which may leave scars, the transmission of disease, or in rare cases death. The consequences for the aggressive dog may be a beating, abandonment, or euthanasia.

What exactly constitutes canine aggression? The types of behaviour that are generally taken as markers for canine aggression are snapping, attempting to bite, and actual biting. An extreme of aggressive behaviour, sometimes referred to as *rage*, occurs when the dog unaccountably becomes highly aggressive, typically towards its owner or members of the household. While evident in several breeds, the phenomenon of rage has become particularly associated with the English Cocker Spaniel (Podberscek & Serpell, 1996).

How may canine aggression be measured? Duffy, Hsu, and Serpell (2008) note four approaches to quantifying aggression, each with its own advantages and disadvantages: (1) statistics on dog bites, (2) inspection of cases presented at behaviour clinics, (3) the opinion of professionals such as veterinarians and dog trainers, and (4) direct behavioural observation and testing.

A study of aggression in 33 breeds of dogs reported by Duffy et al. (2008) revealed that some dogs are generally aggressive, while others, as shown in Table 3.2, are more selective in their targets.

To put these figures in context, Duffy, Hsu, and Serpell note that there is substantial variation *within* breeds indicating that while the propensity for aggression is genetic, environmental influences play a major role in the expression of aggressive behaviour. In a study looking at the behaviour of German Shepherd dogs, Friedrich et al. (2019) found that the way in which the owner managed their dog was significantly important with respect to the dog's behaviour. Thus, factors such as the frequency of training sessions and taking part in dog competitions reduced the likelihood of undesirable behaviours. Podberscek and Serpell (1997) found similar environmental influences at play when comparing high- and low-aggressive English Cocker Spaniels.

Hsu and Sun (2010) used an assessment instrument called the *Canine Behavioral Assessment and Research Questionnaire(C-BARQ)* to look at the characteristics associated with aggression. A survey of 852 dog owners in Taiwan found that their dogs scored highest on aggression to other dogs, then stranger-directed aggression, and finally owner-directed aggression. This pattern of aggressive behaviour is comparable to that found by previous research in America (Duffy et al., 2008) and in The Netherlands (van den Berg et al., 2006). Hsu and Sun make the point that:

> These results probably show a universal trend for dog owners not to tolerate their dogs' aggression toward themselves but to have higher tolerance for aggression toward other dogs. Dog owners are also more tolerant of their dogs showing aggressive responses toward strangers, which is not surprising as they are usually expected to act as guards and encouraged to bark at strangers. (p. 118)

As may be expected, Hsu and Sun found the dog's breed was associated with displays of aggression, while those dogs whose owners used physical punishment were significantly more aggressive. This latter finding resonates with the familiar dictum that "violence begets violence" evident in studies of violence between people (Hollin, 2016; Widom, 1989).

Table 3.2 Dogs most likely to exhibit aggression (from Duffy, Hsu, and Serpell, 2008)

Generally Aggressive: Dachshunds, Chihuahuas, Jack Russell Terriers
Aggressive Towards Other Dogs: Akitas, Siberian Huskies, Pit Bull Terriers
Stranger-Directed Aggression: Dachshunds, Chihuahuas, Doberman Pinschers, Rottweilers, Yorkshire Terriers, Poodles
Owner-Directed Aggression: Basset Hounds, Beagles, Chihuahuas, American Cocker Spaniels, Dachshunds, English Springer Spaniels, Jack Russell Terriers

The second issue concerns where the aggression is targeted: is it towards another animal, typically a dog (sometimes in the same household), or at a stranger, or at its owner? Following the empirical literature, what is known about the individuals, both dogs and owners, and the settings in which the aggressive behaviour occurs?

Aggression towards other dogs

A study by Roll and Unshelm (1997) looked at dogs brought by their owners to a veterinary clinic in Germany. There were 151 owners who saw the vet because their dog had been injured by another dog ("victims") and 55 owners whose dog had caused injuries to other dogs, some several times ("aggressors"). Roll and Unshelm compared the dogs and the owners in the victim group with their counterparts in the aggressor group. They found that most breeds of dog, mainly males, appeared in both the victim and aggressor groups; for example, there were 87 German Shepherds brought to the clinic of which 73 were aggressors and 14 were victims; similarly of the 11 Cocker Spaniels, 3 were aggressors and 8 were victims. The owners said that in the majority of cases both the dogs involved were off the leash, most frequently in a public place, when the fight occurred.

A comparison of the owners revealed several interesting variations across the two groups. The aggressors were mainly owned by men, typically 30–39 years of age, who had selected a specific breed of dog and obtained it from a specialist breeder. These owners had kept dogs for most of their life and they said they had no emotional relationship with their dog. They were likely to be involved in Schutzhund training – a type of training akin to a sport where dog and owner compete to gain ascending levels of proficiency in tasks of stamina and endurance – and to own a dog for security reasons. Within this group, physical force was frequently used to ensure the dog's obedience to their commands. At the time of the fight these owners tended to be passive and then shouted at the dog after the fight.

The owners in the victim group were mainly women, some of an age where they drew their pension, who did not usually select their dog by breed and made the purchase from a friend. These owners keep a dog as a family pet or to ward off loneliness and to provide a feeling of safety. Fewer of this group said they had always had a dog than the owners in the aggressor group. In the main, these owners used more gentle methods of training and after the fight sought to console their dog.

Roll and Unshelm conclude that: "It is clearly shown that not only characteristics of the dogs belonging to the group of victims and aggressors are found (including breed, gender, background, training and housing), but also typical characteristics of the dog owners. Therefore, it is not enough to issue rigid laws for the prevention of potential and real aggression in dogs, based on a breed classification" (p. 242). The use of laws to curtail aggressive dogs is discussed in Chapter 5.

Given the sample in the Roll and Unshelm study, dogs taken to the vet after a fight, is not straightforward to extrapolate directly from these findings to the wider population. Casey, Loftus, Bolster, Richards, and Blackwell (2013) used a different methodology in recruiting a convenience sample of 3,897 dog owners across the UK to answer a questionnaire asking a wide range of questions about themselves, their dog, training methods, and undesirable behaviours including aggression. In all, 871 owners reported that their dog had been aggressive to other dogs while walking and 326 owners reported incidents of aggression towards other dogs in the household. In keeping with Roll and

Unshelm, the owners over 60 years of age were less likely to have a dog that showed aggression; although unlike Roll and Unshelm, there was no difference between male and female owners in ownership of an aggressive dog. When compared with crossbreeds, terriers and pastoral breeds were most likely to show aggression towards dogs outside the household. Those dogs obtained from a rescue centre, as compared with those acquired from a breeder, had the highest risk of aggression to dogs outside the household, with an elevated risk also evident when the dogs came from sources such as pet shops and newspaper adverts.

Aggressive encounters between dogs in public places can be distressing for all concerned but what if the incident involves a guide dog? Brooks, Moxon, and England (2010) considered 100 instances in the UK where a guide dog was attacked by another dog. They estimated that these attacks took place at the rate of more than 3 per month and that 61 of the incidents involved guide dogs, mainly male, that were wearing a harness and working with an owner or trainer. The attacks mainly occurred in public places between 9 o'clock in the morning and 3 o'clock in the afternoon, reflecting when people are most likely to be out and about. In all, 61 of the attacking dogs, of which the bull breeds were most common, were off the lead when the incident took place. In 41 cases, the guide dog needed veterinary attention and in 19 instances there was an injury to the handler or to a member of the public. The attacks were reported to have affected the working performance and behaviour of 45 victim dogs and two dogs had to be withdrawn from working as guide dogs.

Moxon, Whiteside, and England (2016) reported an update of their 2010 figures. They reviewed 629 attacks on guide dogs that took place over the 56 months between June 2010 and February 2015. There was an average of 11.2 attacks per month; with 50 attacks involving two or more aggressors. Brooks, Moxon, and England reflect that this rise may be due either to an increased level of reporting or a real trend (it could also be caused by sampling differences between the studies). As in 2010, almost all of the incidents occurred in a public place; the majority of injured dogs were qualified guide dogs and over one-half were working in harness when attacked. Moxon, Whiteside, and England give the estimated veterinary costs following the attacks as £34,514.30. The attacks incurred further costs in that the dog's working ability was adversely affected in over 40% of cases, while 20 dogs had to be permanently withdrawn from service, 13 of which were working guide dogs, thereby impacting significantly on their owner's mobility and independence as well as their psychological well-being.

Finally, as noted by Casey et al. (2013), aggression can occur when dogs are in the same household. Wrubel, Moon-Fanelli, Maranda, and Dodman (2011) examined the details surrounding 38 pairs of dogs where household aggression, sometimes several incidents a week, had taken place. In 30 cases, the dogs were of the same sex, with females most commonly involved. In 27 cases, there were sufficient details to identify the instigator of the aggression which was typically the younger of the pair (20/27 cases) or a newer addition to the household (19/27). The triggers for fighting, as might be predicted, were competition for owner attention, food, excitement, and disputes over toys and items found by the dogs.

Aggression towards people

A dog's aggression towards people, generally seen in the form of snapping, growling, snarling, and even biting, may be directed towards people it knows, typically the owner and family members, or towards strangers. King, Marston, and Bennett (2012) make the point that while we humans may have changed in the way we perceive our pet dogs, the dogs may not have changed accordingly and so continue to show "undesirable" behaviours. It follows that there would be advantages in being able to assess a dog's shortcomings such as aggression, and its positive attributes such as safety with children and friendliness towards people. The accurate assessment of these traits could inform the buyer's choice of breed, the care and training in dog shelters, and breeding programmes. To this end, several scales have been developed, such as The Dutch Socially Acceptable Behaviour Test (Dalla Villa et al., 2017; van der Borg et al., 2010), to aid assessment of a dog's likely behaviour, including aggressive behaviour (Klausz, Kis, Persa, Miklósi, & Gácsi, 2014). Although, of course, as dogs cannot talk, these assessments inevitably rely on the imperfections of human observation (Wilsson & Sinn, 2012).

The Canine Behavioral Assessment and Research Questionnaire (C-BARQ; Hsu & Serpell, 2003) was developed to provide a systematic means of assessing canine behaviour and temperament. It may also be used to monitor and evaluate the effects of treatment directed at changing problematic behaviours. The C-BARQ is a 68-item questionnaire that assesses a dog's behaviour across 11 domains: these domains, along with a sample item for each, are shown in Table 3.3.

The original C-BARQ has undergone several revisions: the C-BARQ website (https://vetapps.vet.upenn.edu/cbarq/; accessed February 2109) gives full details. The website currently states that the C-BARQ provides an assessment of 14 categories of behaviour as well as "Information on the occurrence of a further 22 miscellaneous behaviour problems ranging from coprophagia to stereotypic spinning/tail-chasing."

Table 3.3 C-BARQ factors and sample items (from Hsu & Serpell, 2003)

Factor 1. *Stranger-directed aggression*
Sample Item: Dog acts aggressively when approached directly by an unfamiliar male while being walked or exercised on a leash.
Factor 2. *Owner-directed aggression*
Sample Item: Dog acts aggressively when verbally corrected or punished by a member of the household.
Factor 3. *Stranger-directed fear*
Sample Item: Dog acts anxious or fearful when approached by an unfamiliar adult male while away from the home.
Factor 4. *Non-social fear*
Sample Item: Dog acts anxious or fearful in response to sudden or loud noises.
Factor 5. *Dog-directed fear or aggression*
Sample Item: Dog acts aggressively when approached directly by an unfamiliar male dog while being walked or exercised on a leash.
Factor 6. *Separation-related behaviour*
Sample Item: Dog displays shaking, shivering, or trembling when left or about to be left on its own.
Factor 7. *Attachment or attention-seeking behaviour*
Sample Item: Displays a strong attachment for a particular member of the household.
Factor 8. *Trainability*
Sample Item: Dog returns immediately when called while off-leash.
Factor 9. *Chasing*
Sample Item: Dog acts aggressively toward cats, squirrels, and other animals entering its yard.
Factor 10. *Excitability*
Sample Item: Dog over-reacts or is excitable when a member of the household return home after a brief absence.
Factor 11. *Pain sensitivity*
Sample Item: Dog acts anxious or fearful when examined or treated by a veterinarian.

Owner-directed aggression

Hsu and Sun (2010) note that dogs with higher C-BARQ scores for owner-directed aggression are older male dogs, typically neutered, and kept outside the house. These dogs were likely to have female owners with no or few other dogs in the household. Bálint, Rieger, Miklósi, and Pongrácz (2017) used a series of assessments to sort dogs into the two groups of "obedient" and "aggressive towards owner." A comparison of the two groups showed that the less-obedient dogs were young, male, and not neutered; the dogs aggressive towards their owner were spayed/neutered in keeping with Hsu and Sun (2010).

Stranger-directed aggression

Several studies have used the C-BARQ to examine stranger-directed aggression. A study conducted in Twain used a Chinese revision of the C-BARQ containing 103 items, grouped into seven domains, to look at aggression in pet dogs (Hsu & Sun, 2010). It was found that higher scores on stranger-directed aggression were associated with dogs that were acquired either as puppies or specifically for the purpose of guarding property, living in rural areas, and in houses with yard space and with a greater number of household members. A study by van den Berg, Heuven, van den Berg, Duffy, and Serpell (2010) used scores on the stranger-directed domain from the C-BARQ to look at aggression towards strangers in a sample of 1,000 dogs consisting of 333 German Shepherds, 224 Golden Retrievers, and 443 Labrador Retrievers. They found that "German Shepherds have significantly higher scores for aggression than the Labradors, and that the Labradors in turn have significantly higher scores than the Golden Retrievers" (p. 139).

Flint, Coe, Serpell, Pearl, and Niel (2017) examined C-BARQ data from on a sample of 11,240 dogs categorised as aggressive towards strangers, with 1,125 of these dogs labelled as "severely aggressive." When compared to dogs with no fear of strangers, the dogs seen as severely aggressive were male, rated by their owner as mildly or severely fearful of strangers, and were fearful in non-social situations. The breed group and where the dog was acquired also had an association with severe aggression. Thus, hounds such as greyhounds and beagles were unlikely to be aggressive to strangers, while mixed breeds were more likely to show this form of aggression.

For many people, the anxiety triggered by an aggressive dog is that it will bite someone, a fear that may be exaggerated if the potential victim is a child. A body of knowledge has accumulated which gives insights into the factors leading up to the bite, its psychological and physical effects, and the ensuing costs.

Biting

What, exactly, is a dog bite? Oxley, Christley, and Westgarth (2019) make the point that while we may talk about dog bites using distinctions such as a "nip" or a "play bite," the study of bites needs more precise definitions. Oxley, Christley, and Westgarth conducted a survey using an online questionnaire asking people over the age of 18 years to give details of their most recent dog bite. There were 484 responses mainly from people living in England, most were female (84.8%) with the most common age groups of 45–54 years (24.9%) and 35–44 years (24.4%). The majority of respondents said they currently (82.6%) or previously owned a dog (87.7%). In their most recent incident, the majority of respondents (86%) stated they were bitten once. As can be seen from Table 3.4, respondents gave a range of responses when they said what would qualify as a bite.

Alongside the diversity of views on what constitutes a dog bite, opinion was also divided as to role of the dog's intentions. Just over 40% of respondents said that if a dog did not intend to bite, as say during play, then it would not qualify as a bite. Finally, over one-half of the respondents said it would be unlikely that they would seek medical attention if the dog belonged to a friend or family member or if it was their own dog.

Oxley et al. (2019) illustrate the difficulties in researching dog bites. There are marked disagreements between people on what actually defines a dog bite. This definitional issue makes it difficult to compare across studies as researchers and their respondents may have different operational definitions of a "bite." In some instances, the injured person is reluctant to seek medical attention. It follows that research which uses samples drawn from those seeking medical attention is selective and so potentially biased. These methodological issues should be seen as germane across research in this area and held in mind when reading the evidence.

In the interests of not giving a dog a bad name, dogs are not the only animal that bites or causes injury. Langley, Mack, Haileyesus, Proescholdbell, and Annest (2014) reported a study of people treated in an Emergency Department (ED) in America between 2001 and 2010. They found that an estimated 10.1 million people, about one-third of whom were aged 14 years or under, visited an ED for sting injuries and non-fatal bites not caused by a dog. The main aggressors were arachnids (scorpions and toxic spiders) and insects (principally bedbugs, bees, and wasps), while the main mammalian culprits were cats and rodents. Another American study by Forrester, Weiser, and Forrester (2018) reviewed the mortality data for the period 2008 to 2015 concerning 1,610 deaths caused by venomous and non-venomous animals. The deaths caused by venomous animals were principally due to hornets, wasps, and bees (478 deaths); venomous arthropods, such as scorpions (84 deaths); and snakes and lizards (83 deaths). Of the deaths by non-venomous animals, 272 were attributed to dogs and 576 to other mammals which included farm livestock, principally horses and cattle. The most deaths (813) were in the 35–64-year age group, with 172 deaths in the 0–9-year-old group. Finally, an Indian study by Wani and Sabah (2018) reported that while almost 90% of bite-related admissions to a medical centre involved a dog, the other biting animals included bears, cats, horses, and monkeys.

Table 3.4 Percentage of respondents ($N = 484$) agreeing what constitutes a dog bite (from Oxley et al., 2019)

Obnoxious or aggressive behaviour but no contact by teeth:	3.4%
Dog only made contact with clothing:	45.5%
Skin contact by teeth but no skin puncture or bruising:	62.8%
Skin contact by teeth and bruising but no skin puncture:	81.3%
One to four punctures from a single bite with no puncture deeper than half the length of the dog's canine teeth:	91.9%
One to four punctures from a single bite with at least one puncture deeper than half the length of the dog's canine teeth:	90.9%
Multiple bite incident with at least two deep bites:	88.2%

What's in a bite?

The widespread concern with dog bites is reflected in the volume of research from around the globe looking at their incidence and consequences. An example of this type of research is provided by De Keuster, Lamoureux, and Kahn (2006) who carried out a telephone survey in Belgium based on 8,000 randomly selected home numbers from which 1,184 families with at least one child under 15 years of age finally participated. Of the 1,184 families who took part, 26 reported a dog bite to a child, an annual rate of 2.2%. When a child had been bitten in the preceding 12 months, some additional questions were asked about the incident. Of the 26 children, 10 saw a general practitioner, 5 went to a hospital emergency department, and 1 child was admitted to hospital; thus, 10 bites went unreported.

In a second part to their study De Keuster, Lamoureux, and Kahn gathered prospective data from six hospital emergency departments. The paediatricians collected standardised information on all child bite victims younger than 16 years admitted for treatment. The child or an adult or caregiver answered a questionnaire concerning the victim, the dog, and details of the incident. A total of 100 completed questionnaires revealed that on average, three children who had been bitten were seen per month in the emergency departments. How does this rate of admission compare with other childhood accidents? De Keuster, Lamoureux, and Kahn note that in the same period and the same hospital emergency facilities there was an average of 11.5 admissions following a road traffic accident and 10 with burns following a fire. In all, the dog bites were the reason for one-quarter of 1 percent of children brought to the hospital.

With regard to the 100 bitten children, 65 were about 4 years of age, were bitten at home without an adult present, and 61 knew the dog; 35 children, about 9 years old, of which just 10 knew the dog, were bitten in a public place. The child had probably triggered the incident in 56 of the 65 home incidents and 11 of the bites in a public place. The breeds of dog most often involved were German Shepherds (28 times), which were also most often involved in multiple bites [15/25]); Rottweilers (11); and Labradors (9). In the large majority of cases, medical assistance was sought within a few hours of the bite. At follow-up, two children, one bitten on the face at home, the other bitten on the leg when alone in a public place, were receiving treatment from a psychologist.

The concerns raised in the Belgium study can be found in the global literature with a focus on children who are bitten. Thus, a study conducted in the Czech Republic by Náhlík, Baranyiová, and Tyrlík (2011) gained details of dog bite incidents with 92 children aged 4 to 11 years by asking the children to complete a questionnaire. In the majority of cases, these incidents occurred when the child was playing with the dog, or if the dog tried to take something from the child, or the child caused pain to the animal. The pattern of an interaction between a person and a dog which precedes the bite, sometimes necessitating medical treatment, is evident in countries as diverse, among others, as Australia (Rajshekar et al., 2017), Cambodia (Ponsich, Goutard, Sorn, & Tarantola, 2016), China (Shen, Lib, Xiang, Lub, & Schwebel, 2014), the Czech Republic (Náhlík et al., 2011), France (Sarcey, Ricard, Thelot, & Beata, 2017), India (Ovais, Adil, Darzi, & Beenish, 2017), Iran (Kassiri, Kassiri, Lotfi, Shahkarami, & Hosseini, 2016), Ireland (Ó Súilleabháin, 2015), Israel (Cohen-Manheim, Siman-Tov, Radomislensky, Peleg, & Israel Trauma Group, 2018), Italy (Alberghina, Virga, Buffa, & Panzera, 2017), Korea (Park et al., 2019), New Zealand (Wake, Minot, Stafford, & Perry, 2009), Nigeria (Omoke & Onyemaechi, 2018), Serbia (Vučinić & Vučićević,

2019), Spain (Rosado, García-Belenguer, León, & Palacio, 2009), Switzerland (Horisberger, Stärk, Rüfenacht, Pillonel, & Steiger, 2004), Trinidad (Georges & Adesiyun, 2008), USA (Loder, 2019), and the UK (Westgarth, Brooke, & Christley, 2018).

It is not surprising that, as Polo, Calderón, Clothier, and de Casssia Maria Garcia (2015) note, The World Health Organisation has called dog aggression a public health problem.

A similar pattern to the large-scale studies emerges when researchers focus on small communities. West and Rouen (2019) reported an analysis of dog bite injuries, using data from 1st January 2006 to 31st December 2011 gathered from a clinical file audit at Primary Health Care Clinics in three remote Indigenous communities within Far North Queensland, Australia. In the period covered by the study, 201 people presented with 229 dog bites. West and Rouen calculate that the figures equate to an overall incidence rate of 16.5 per 1,000 of the Indigenous community's population.

Bjork et al. (2013) looked at dog bites requiring hospital treatment among Alaska Native and American Indian children (defined as under 20 years of age) between 2001 and 2008. The average annual dog bite hospitalisation rate was higher for both populations (Alaska Native, 6.1 per 100,000; American Indian 5.3 per 100,000) than for the general child population in the USA (3.1 per 100,000). Chang, McMahon, Hennon, LaPorte, and Coben (1997) used a capture-recapture methodology to estimate the incidence of dog bites in the American city of Pittsburgh. They gathered data from several sources, including both hospital and police records, and noted that in 1993, 790 dog bite injuries were reported, an incidence rate of about 2.14 per 100,000 of the city's population. This methodology allows an estimate to be made of unreported bites which Chang et al. gave as an estimated annual incidence rate of 5.89 per 100,000 of the population. Indeed, as Beck and Jones (1985) point out, not only are many bites unreported, accounts of what happened during the bite are subject to the foibles of human memory.

Given the intrinsic methodological issues, there is a remarkable degree of concordance across the global literature with respect to dog bites. First, anyone of any age can be bitten but it is preschool-age children, around 2–5 years of age, who are at the greatest risk of a dog bite; second, bites can result in major injury, particularly to the head and neck region; third, while death is rare, the injuries from the bite may well require medical treatment.

When compared to other common non-medical causes of childhood death such as road traffic and household accidents (Hon & Leung, 2010) the overall number of fatalities due to dog bites is low. This point is emphasised by Raghavan (2008), who reported that in Canada between 1990 and 2007 there were just 28 known fatalities from dog bites. The vulnerability is children is emphasised as 24 of the 28 fatalities were children under the age of 12 years. Messam, Kass, Chome, and Hart (2018) investigated factors associated with bites when the child (aged 5–15 years) and dog lived in the same home. Messam et al. found that younger children, particularly boys, were at greatest risk of a bite. The dogs most likely to bite were not neutered, lived in a home with no outside space, were regularly allowed inside the house, and were habitually allowed to sleep in a family member's bedroom. Guy et al. (2001) also commented that allowing a dog to sleep on a family member's bed was a feature of homes where a bite had occurred.

Children are not the only overtly vulnerable group with regard to dog bites. A large-scale study by Yeh et al. (2012) carried out using national data in Taiwan looked the experience of dog bites by people with a mental disorder. They found that a that a mental disorder, both psychotic and non-psychotic, heightened the risk for dog bites and associated injuries. However, the anxiety, dissociative, and somatoform disorders had the greatest risk for dog bites. The explanation for this association is uncertain: it may be a result of the person taking undue risks, perhaps because of their misperception of the situation, or a combination of mental disorder and sociodemographic status that leads to increased contact with dogs.

While it is important to retain a sense of perspective, injuries and (rarely) death are caused by dog bites. The widespread concern about the costs, both personal and in terms of health care, of a bite has produced a considerable empirical literature over the years. The results of this research can be summarised using Skinner's three-term contingency that considers the *antecedent* conditions to the bite, the actual *behaviour*, and the *consequences*.

Antecedents to biting

There are two dimensions to consider in looking at antecedents to the bite. First, there are *distal* antecedents, such as factors in the dog's upbringing; and there are the *proximal* antecedents, say taking away a toy, that immediately precede the bite. The antecedents to a bite may be thought of as risk factors for biting: the presence of a risk factor does not make it inevitable that the behaviour will occur, rather that the likelihood of it happening is heightened. The benefit of identifying risk factors lies in the information they provide for the development of interventions to prevent biting (see Chapter 4).

Distal antecedents

As discussed above, background factors, such as being a rescue dog, increase the chances of aggression, while some breeds of dog, such as the German Shephard, are more likely than others to be aggressive. The importance of a dog's learning history with regard to its training and everyday routines is seen in a study by O'Sullivan, Jones, O'Sullivan, and Hanlon (2008) of the histories of 100 dogs who had bitten a person. The analyses revealed: "A significant association between a history of aggressive behaviour and the combined variables of a dog not being socialised with other dogs, not responding to the command 'sit', not being socialised with children, displaying fearful reactions to specific stimuli, being punished verbally/physically by the owner and being fed directly from the family table recently" (pp. 153–154). While not all the variables had the same explanatory power, the point is made that combinations of factors in a dog's history can increase the risk of biting.

Guy et al. (2001) presented a case series of 227 dogs that had bitten a person who either lived in same household or was a regular visitor to the house. The breed most often found in this series was the Labrador Retriever, a breed not particularly noted for its aggressiveness. The majority of bite victims were adults, which runs contrary to the more usual child victims. These findings indicate the importance of context such that the risk factors for biting may vary from public to private environments. On a grander scale, in considering canine aggression it is important to understand the national context. A Chinese study by Shen et al. (2012) noted three areas where the place of dogs in everyday life in some parts of China differs markedly from Europe and America. First, the sheer number of dogs is extraordinary by western standards: Shen et al. give estimates of between 75 and 200 million dogs in China (by comparison the estimated canine population of the UK is 8.5 million). Second, as many men leave their rural homes to work in the cities, the families they leave behind obtain a dog for protection. However, many of these dogs escape and form packs, moving about the countryside, reproducing freely. As many children in rural China walk to school so a context is created in which the children become potential targets for the stray dogs. Third, the hazard presented by these dogs is compounded by the facts that many dogs, particularly in southern China, have rabies: "From 1996 to 2007, China experienced a human rabies incidence increase of over 2000%, from 159 to over 3300 cases annually" (Shen et al., p. 22).

Proximal antecedents

What happens in the short period of time preceding the bite? Arhant, Beetz, and Troxler (2017) reported a survey of 402 families with dogs who lived in Austria and Germany. The purpose of this study was to provide normative data, as reported by caregivers (mainly mothers), regarding the nature of interactions between the child and dog. The most frequently observed interactions between the children, average age 2.5 years, and the dogs were caring in nature: these interactions included patting the dog's body and head, talking to the dog, and moving towards or following the dog; the more problematic behaviours with regard to the risk of biting, such as kissing and hugging, were less common. A child trying to interact with a resting dog, say by trying to wake it up, was unusual, although some children did lie down close to a resting dog.

A few children tried to interact with the dog when it was feeding, either by petting the dog or moving its feeding bowl. Some children attempted to take back their toys from the dog. There were also interactions between the child and dog which the animal found to be aversive: these included the child yelling, reprimanding the dog and, less often, dressing the dog, using the dog in a game, and lifting the dog. It was highly unusual for a child to hurt the dog: there were a small number of cases of the child deliberately hitting, kicking, or throwing objects at the dog; more often, the child would pull at the dog's ears or tail, ride on the dog's back, or accidentally step on the dog. In an analysis of facial dog bites to children, Chen, Neumeier, Davies, and Durairaj (2013) noted that the child provoked the dog in 164 of the 308 documented cases. The most common provocations were aggressive petting and play, startling the dog, and the child falling or stepping onto the dog. There are several studies which add weight to the combination of a home location with the child petting or playing with the dog when the bite occurred (e.g., Abraham & Czerwinski, 2019; Messam, Kass, Chome, & Hart, 2012).

Rezac, Rezac, and Slama (2015) looked at the behaviour of 132 people, 92 of whom were under 18 years of age, immediately prior to a facial bite. Not surprisingly, the incidents occurred most frequently when the person bent over the dog or put their face close to the dog's face. In most cases, the dog was off the leash and the incident took place on the dog owner's property. The location of the bite was independent of the person's age and gender and the size of dog. Owczarczak-Garstecka, Watkins, Christley, and Westgarth (2018) also noted that alongside standing or leaning over a dog, the level of tactile contact, such as petting, hugging, hitting, and restraining, with the dog increased in the 20 seconds preceding the bite.

Humans communicate with various modes of non-verbal communication, such as facial expression, eye contact, and gesture (Argyle, 1988). Dogs communicate in the same way using body posture, such as ear position (e.g., down, erect), tail position, licking, and growling, to communicate messages regarding their emotional state. If these signals are misunderstood or ignored and the person acts inappropriately, the dog may escalate its behaviour leading to unwanted consequences (McGreevy, Henshall, Starling, McLean, & Boakes, 2014). Shepherd (2009) describes a "ladder of aggression" showing the escalating steps a dog takes in signalling its response to stress or a threat. At the lowest rung of the ladder, the dog blinks, yawns, and nose licks, then climbs to standing crouched and tail tucked under its body, escalating to growling and snapping, before biting.

Not all children who have contact with a dog are bitten, leading to the question of whether there are there any characteristics of the child associated with an increased risk of biting. Davis, Schwebel, Morrongiello, Stewart, and Bell (2012) observed 88 children aged from 3.5 to 6 years in their interactions with a dog. The parent-report Children's Behavioral Questionnaire was used to assess four dimensions – impulsivity, inhibitory control, approach, and shyness – of the child's temperament. They found that the less shy children took greater risks with the dog, which was unchanged when controlling for the dog's characteristics.

Biting behaviour

There are two aspects to a bite to consider: (i) the location of bite and (ii) the severity of the bite in terms of physical injury.

Location of the bite

Rezac et al. (2015) found that the bites were mainly to the central area of the face, the nose, and lips, and less often to facial extremities, the chin, cheek, around the eyes, and the forehead. The location of the bite was not related to the person's age and gender or the size of dog. As the majority of bites took place when the person was bending over the dog, the location of the bites is entirely understandable.

Sarcey et al. (2017) reported that for children, rather than adults, the bite was more often to the head and neck. In adults, the bite frequently came about in the context of attempting to separate two fighting dogs; for children, the bite occurred in an interaction with a familiar dog. Oxley, Christley, and Westgarth (2018) found that when the dog approached the person, the bite was to lower body parts, such as legs and ankles; conversely, when the person approached the dog, the bites were to the upper body including arms, hands, and face. In accord with the literature, they found that an interaction between the person and the dog – such as playing, restraining, or stroking – was a frequent context for a bite.

Severity of the bite

Given the issues in deciding whether a bite has taken place (Oxley et al., 2019), it is not an easy task to classify the severity of dog bites. Lackmann, Draf, Isselstein, and Töllner (1992) differentiated five levels of facial bite injury to children according to the need for surgery: (i) superficial injury; (ii) deep injury involving muscle damage; (iii) deep injury involving muscle and tissue defect; (iv) injuries as at (iii) plus vascular and/or nerve injury; (v) injuries as at (iii) plus bone involvement and/or organ defect. This approach to understanding bite severity according to the level of physical harm continues to inform the field. Thus, for example, O'Brien, Andre, Robinson, Squires, and Tollefson (2015) developed a Dog Bite Complication Index based on level of physical damage.

Table 3.5 Dog Bite Severity Ratings (after Dunbar, n.d.)

Level 1. Aggression without biting.
Level 2. Teeth make contact with skin without a puncture; may be slight bleeding.
Level 3. A single bite with one to four punctures wounds no deeper than one-half the length of the dog's canine teeth.
Level 4. As in Level 3, with at least one puncture deeper than one-half the length of the dog's canine teeth. There may be bruising around the wound caused by the dog holding on or sideways lacerations if the dog held on and shook its head.
Level 5. Multiple bites with at least two Level 4 bites or a multiple-attack incident with at least one Level 4 bite in each.
Level 6. Victim dead.

In formulating The Dog Bite Severity Ratings, shown in Table 3.5, Dunbar (n.d.) took a different approach in classifying different types of bite according to particulars such as the depth of the bite, the number of puncture wounds, along with the degree of physical harm.

The bites are not equally distributed across the six levels. Dunbar suggests that the more minor types of bites at Levels 1 and 2 account for over 99% of incidents. In these incidents, the dog may be fearful, even out of control, but is not dangerous.

Owczarczak-Garstecka et al. (2018) used a range of variables – the context in which the bite occurred, the duration of the incident, size of dog, victim's age and sex, anatomical location of the bite, and whether victim or dog initiated the interaction – to formulate a bite severity score. This severity score rose according to several aspects of the incident including the duration of the incident and when there were bites to multiple locations rather than just the limbs or face. The adult victims experienced more severe bites than the child victims, while bites to infants were more severe than those suffered by children. Sarcey et al. (2017) also reported that the lesions resulting from the bite were more severe in adults than in children. The bites were more numerous and more severe when the victim knew the dog.

Essig, Sheehan, Rikhie, Elmaraghy, and Christophel (2019) carried out a systematic review and meta-analysis of 43 studies of facial injury resulting from dog bites. Essig et al. were primarily interested in the risk of biting from various breeds and constructed a means to assess the severity of the bite according to the level of surgery required. Thus, the severity of the injury was classified according to a 6-point measure of tissue damage formed from an aggregation of wound size and tissue avulsion. The degree of surgery needed for each of the six levels of tissue damage – ranging from simple closure, through layered closure, to complex closure requiring tissue rearrangement, and finally tissue loss and fractures – pointed to the severity of the bite. The most severe injuries were caused by Pit Bull terriers and mixed breed dogs.

The severity of a dog bite, clearly of medical concern, can be viewed as a combination of the particulars of the bite and the associated level of physical damage. The next issue is the fate of the bitten and the biter.

Consequences of biting

In their one-month follow-up of 292 people who had sought emergency treatment for dog bites, Sarcey et al. (2017) found that 114 people reported some consequence. These consequences ranged from the *aesthetic* in nature, such as the presence of scarring (91 of 114), the *physical* such as a loss of mobility (17 of 114), and the *psychological* including nightmares and a fear of dogs (6 of 114). Sarcey et al. noted that women reported more consequences than men and adults more than children. As with the antecedents to bite, there are proximal and distal consequences to consider.

Proximal consequences

The immediate aftermath of a dog bite has two principal features: (1) the physical damage caused by the bite and (2) the short-term psychological effects. In the short term, there is the physical pain caused by the bite which, as discussed above, can vary in severity. Oxley et al. (2018) found that most injuries resulting from dog bites did not require medical treatment. Nonetheless, given individual differences in pain perception (Nielsen, Staud, & Price, 2009), the person's experience of being bitten will range from mild to extreme pain. It can be taken that multiple bites, perhaps involving several dogs, will be highly painful, with immediate death a possibility, particularly so if the victim is a very young child (Golinko, Arslanian, & Williams, 2017).

When the injuries are serious, immediate medical attention is needed which may include surgery. Garvey, Twitchell, Ragar, Egan, and Jamshidi (2015) reviewed 282 cases admitted to Phoenix Children's Hospital, Arizona, following a dog bite. They found that the most common injuries (231/282) were skin injuries including punctures, avulsions, and simple or complex lacerations. The most severe injuries were to the face (16/282), body extremities and pelvis (16/282), head and neck (13/282), abdomen (5/282), male genitalia (4), and chest (1/282). Some dogs have a powerful bite and 25 cases had fractures, some to the skull. Almost all of the injuries (259/282) required surgery including four amputations (4/282) (three fingers and an ear). The extent of the problem of dog bites has necessitated advances in surgical techniques (Saadi, Oberman, & Lighthall, 2018). Alongside the physical damage, there may be other heath hazards, such as rabies and severe bacterial infections, which require medical attention (e.g., Abrahamian & Goldstein, 2011; Babazadeh et al., 2016).

While the immediate concern is with the severity of the physical injury and the need for medical treatment, the psychological effects of the bite may be that the victim experiences feelings of panic and fear as well as anger at the dog or its owner. However, it is in the longer term that the psychological consequences of being bitten may begin to be felt.

Distal consequences

Dhillon, Hoopes, and Epp (2018) review decades of research into the consequences of a dog bite from which can be distilled four distinct distal consequences: (1) the effects of the *physical* damage, (2) the *psychological* repercussions, (3) the *financial* costs of the incident, and (4) the wider *public and political* ramifications. Sarcey et al. (2017) highlight the scale of the issue in a 1-month follow-up of 292 people who had accessed medical services after a dog bite. The 114 cases reported a sequelae to the bite, most commonly aesthetic due to scarring or disfigurement (91 of 114), physical, (17 of 114), and psychological (6 of 114). In addition, there may be consequences for the dog: it is not uncommon for the dog to be physically punished or put down.

Physical consequences

The enduring physical effects reported by Sarcey et al. included pain, loss of mobility, a physical handicap (motor or sensitive), and a social handicap associated with physical scarring. Some of these physical consequences, most notably scarring, are inextricably linked with psychological and social functioning. There is, for example, a strong association between facial scarring and psychological distress both generally (Hull et al., 2003) and with specific disorders such as anxiety and depression (Gibson, Ackling, Bisson, Dobbs, & Whitaker, 2018).

Psychological consequences

There is a range of psychological sequelae to a bite including nightmares, anxiety, and affected mood. Stacey et al. reported that about one-third of bite victims said they had a resultant fear of dogs. This fear can develop into a specific phobia (cynophobia), which can be socially disabling; it is difficult to be in some public places without encountering a dog.

In extremis, the fear can be manifest as post-traumatic stress disorder (PTSD), the symptoms of which, summarised in Table 3.6, are defined in DSM-V (American Psychiatric Association, 2013). An individual may experience some of these symptoms, *partial PTSD* or, in *full PTSD*, one or more symptoms from each of the five clusters. In full PTSD, the distress or impairment brought about by the symptoms must have persisted for at least 1 month and must not be related to illness, medication, or substance use.

Table 3.6 DSM-V symptom clusters for PTSD

1. **Stressor**
    Direct exposure to trauma
    Witnessing a trauma
    Exposure to trauma (e.g., police, firefighter)
    Finding out that someone close experienced the trauma
2. **Intrusion symptoms**
    The individual exposed to the trauma re-experiences the incident through any of:
    Flashbacks
    Nightmares
    Distressing and intense memories
    Distress or physical reactions after a sensory reminder of the traumatic incident
3. **Unpleasant changes to mood or thoughts**
    Blaming self or others for the trauma
    Loss of interest in previously enjoyed activities
    Negative feelings about self and the world
    Inability to remember the trauma clearly
    Finding it difficult to be positive
    Feeling isolated
4. **Avoidance**
    Any one of:
    Avoiding external reminders of what happened
    Avoiding trauma-related thoughts or emotions, maybe through drugs or alcohol
5. **Changes in reactivity**
    Any two of:
    Aggression or irritability
    Hypervigilance and hyper-awareness
    Difficulty concentrating
    Difficulty sleeping
    Heightened startle response
    Engaging in destructive or risky behaviour
    Difficulty sleeping or staying asleep

Children may also experience PTSD, displaying a range of symptoms similar to adults. In addition, children are also likely to wet their bed, be unusually anxious about separation from their parent or carer, and may try to re-create the traumatic event in their play. The psychological effects do not have to be restricted to anxiety and affective disorders. Anyfantakis, Botzakis, Mplevrakis, Symvoulakis, and Arbiros (2009) present the case of a 4-year-old girl who, after extensive lesions due to a dog bite, displayed selective mutism. Peters, Sottiaux, Appelboom, and Kahn (2004) reviewed 22 cases of children under the age of 16 years admitted to the Emergency Department of the University Children's Hospital in Brussels following a dog bite. They found that 12 of the children had had PTSD symptoms for more than a month, including flashbacks and fear of dogs, and 5 children had full PTSD. In some cases, the child would not leave the house alone and others became shy and aggressive. Peters et al. add that: "Of the 22 child victims of dog bites ... none received psychological support" (p. 121).

Alongside psychological distress, the presence of facial scars can affect the individual's body image leading to problems in social interaction (De Sousa, 2010; Rumsey & Harcourt, 2004). These difficulties may precipitate a range of complications affecting the person's family life, social activities, and employment (Macgregor, 1990).

Financial consequences

The medical care necessary for a dog bite, increasing as with the severity of the bite, is the most immediate financial cost. It is not a simple matter to estimate the cost of health care – particularly with regard to making international comparisons and allowing for changes in cost over time – given the multiplicity of factors involved. The difficulties in estimating medical costs are perfectly illustrated in a study by Babazadeh et al. (2016) of the costs of rabies vaccination for animal bites. Babazadeh et al. explain that it is possible to be precise about the financial cost of the vaccine but there are additional costs to consider such as the expense of storing and transporting the vaccine and the costs of syringes. Yet further, there are also professional costs in the time of the medical and paramedical staff, hospital administrators, and other support staff. There may be further non-medical costs: the injured person may be absent from work with costs for their employer and the need for sickness or disability benefit.

If it is far from simple to estimate the cost of a vaccination then the complexity increases exponentially when surgery is necessary. It is fair to conclude that the cost of bites to both the individual and the state will be extremely high.

Social consequences

The ubiquity of injuries from dog bites can lead to public disquiet about the presence of dogs. This situation may lead to dog-friendly innovations such as dog parks which benefit both owners and dogs, or to dog-hostile reactions, as will be seen in Chapter 5, in the publicity and legislation surrounding dangerous dogs.

There are several aspects of our behaviour that can be interpreted as high risk with regard to endangering a bite. These risky behaviours include seeking very close proximity to the dog, disturbing the dog while feeding, and causing the dog distress. As will be seen, this knowledge can be put to use in the design and implementation of strategies to reduce the damage caused by biting.

References

Abraham, J. T., & Czerwinski, M. (2019). Pediatric dog bite injuries in central Texas. *Journal of Pediatric Surgery, 54,* 1416–1420.
Abrahamian, F. M., & Goldstein, E. J. C. (2011). Microbiology of animal bite wound infections. *Clinical Microbiology Reviews, 24,* 231–246.
Alberghina, D., Virga, A., Buffa, S. P., & Panzera, M. (2017). Incidence and characteristics of hospitalizations after dog's bite injuries in Sicily (Italy) between 2012–2015. *Veterinaria Italiana, 53,* 315–320.
Amat, M., & Manteca, X. (2019). Common feline problem behaviours: Owner-directed aggression. *Journal of Feline Medicine and Surgery, 21,* 245–255.
American Psychiatric Association. (2013). *Diagnostic and Statistical Manual of Mental Disorders (DSM–5)*. Washington, DC: Author.
Anyfantakis, D., Botzakis, E., Mplevrakis, E., Symvoulakis, E. K., & Arbiros, I. (2009). Selective mutism due to a dog bite trauma in a 4-year-old girl: A case report. *Journal of Medical Case Reports, 3,* 100–102.
Argyle, M. (1988). *Bodily communication*. London: Methuen.
Arhant, C., Beetz, A. M., & Troxler, J. (2017). Caregiver reports of interactions between children up to 6 years and their family dog: Implications for dog bite prevention. *Frontiers of Veterinary Science, 4,* 130.
Asp, H. E., Fikse, W. F., Nilsson, K., & Strandberg, E. (2015). Breed differences in everyday behaviour of dogs. *Applied Animal Behaviour Science, 169,* 69–77.
Babazadeh, T., Nikbakhat, H. A., Daemi, A., Yegane-kasgari, M., Ghaffari-fam, M., & Banaye-Jeddi, M. (2016). Epidemiology of acute animal bite and the direct cost of rabies vaccination. *Journal of Acute Disease, 5,* 488–492.
Bálint, A., Rieger, G., Miklósi, Á., & Pongrácz, P. (2017). Assessment of owner-directed aggressive behavioural tendencies of dogs in situations of possession and manipulation. *Royal Society Open Science, 4,* 171040.
Beck, A. M., & Jones, B. A. (1985). Unreported dog bites in children. *Public Health Reports, 100,* 315–321.
Bjork, A., Holman, R. C., Callinan, L. S., Hennessy, T. W., Cheek, J. E., & McQuiston, J. H. (2013). Dog bite injuries among American Indian and Alaska Native children. *Journal of Pediatrics, 162,* 1270–1275.
Brooks, A., Moxon, R., & England, G. C. (2010). Incidence and impact of dog attacks on guide dogs in the UK. *The Veterinary Record, 166,* 778–781.
Cannas, S., Talamonti, Z., Mazzola, S., Minero, M., Picciolini, A., & Palestrini, C. (2018). Factors associated with dog behavioral problems referred to a behavior clinic. *Journal of Veterinary Behavior, 24,* 42–47.
Casey, R. A., Loftus, B., Bolster, C., Richards, G. J., & Blackwell, E. J. (2013). Inter-dog aggression in a UK owner survey: Prevalence, co-occurrence in different contexts and risk factors. *The Veterinary Record, 172,* 127.
Chang, Y. F., McMahon, J. E., Hennon, D. L., LaPorte, R. E., & Coben, J. H. (1997). Dog bite incidence in the city of Pittsburgh: A capture-recapture approach. *American Journal of Public Health, 87,* 1703–1705.
Charles, N. (2017). Written and spoken words: Representations of animals and intimacy. *Sociological Review, 65,* 117–133.
Chen, H. H., Neumeier, A. T., Davies, B. W., & Durairaj, V. D. (2013). Analysis of pediatric facial dog bites. *Craniomaxillofac Trauma Reconstruction, 6,* 225–232.
Cohen-Manheim, I., Siman-Tov, M., Radomislensky, I., & Peleg, K., Israel Trauma Group. (2018). Epidemiology of hospitalizations due to dog bite injuries in Israel, 2009–2016. *Injury, International Journal of the Care of the Injured, 49,* 2167–2173.
Col, R., Day, C., & Phillips, C. J. (2016). An epidemiological analysis of dog behavior problems presented to an Australian behavior clinic, with associated risk factors. *Journal of Veterinary Behavior: Clinical Applications and Research, 15,* 1–11.

Dalla Villa, P., Barnard, S., Di Nardo, A., Iannetti, L., Vulpiani, M. P., Trentini, R., Serpell, J. A., & Siracus, C. (2017). Validation of the Socially Acceptable Behaviour (SAB) test in a Central-Italy pet dog population. *Veterinaria Italiana*, *53*, 61–70.

Davis, A. L., Schwebel, D. C., Morrongiello, B. A., Stewart, J., & Bell, M. (2012). Dog bite risk: An assessment of child temperament and child-dog interactions. *International Journal of Environmental Research and Public Health*, *9*, 3002–3013.

De Keuster, T., Lamoureux, J., & Kahn, A. (2006). Epidemiology of dog bites: A Belgian experience of canine behaviour and public health concerns. *Veterinary Journal*, *172*, 482–487.

De Sousa, A. (2010). Psychological issues in acquired facial trauma. *Indian Journal of Plastic Surgery*, *43*, 200–205.

Dhillon, J., Hoopes, J., & Epp, T. (2018). Scoping decades of dog evidence: A scoping review of dog bite-related sequelae. *Canadian Journal of Public Health*, *110*, 364–375.

Duffy, D. L., Hsu, Y., & Serpell, J. A. (2008). Breed differences in canine aggression. *Applied Animal Behaviour Science*, *114*, 441–460.

Dunbar, I. (n. d.) The Bite Scale. An objective assessment of the severity of dog bites based on evaluation of wound pathology. Available at http://www.dogtalk.com/BiteAssessmentScalesDunbarDTMRoss.pdf (accessed 9 April 2019).

Essig, G. F., Jr., Sheehan, C., Rikhie, S., Elmaraghy, C. A., & Christophel, J. J. (2019). Dog bite injuries to the face: Is there risk with breed ownership? A systematic review with meta-analysis. *International Journal of Pediatric Otorhinolaryngology*, *117*, 182–188.

Flint, H. E., Coe, J. B., Serpell, J. A., Pearl, D. L., & Niel, L. (2017). Risk factors associated with stranger-directed aggression in domestic dogs. *Applied Animal Behaviour Science*, *197*, 45–54.

Forrester, J. A., Weiser, T. G., & Forrester, J. D. (2018). An update on fatalities due to venomous and nonvenomous animals in the United States (2008–2015). *Wilderness & Environmental Medicine*, *29*, 36–44.

Friedrich, J., Arvelius, P., Strandberg, E., Polgar, Z., Wiener, P., & Haskell, M. J. (2019). The interaction between behavioural traits and demographic and management factors in German Shepherd dogs. *Applied Animal Behaviour Science*, *211*, 67–76.

Garvey, E. M., Twitchell, D. K., Ragar, R., Egan, J. C., & Jamshidi, R. (2015). Morbidity of pediatric dog bites: A case series at a level one pediatric trauma center. *Journal of Pediatric Surgery*, *50*, 343–346.

Georges, K., & Adesiyun, A. (2008). An investigation into the prevalence of dog bites to primary school children in Trinidad. *BMC Public Health*, *8*, 85.

Gibson, J. A. G., Ackling, E., Bisson, J. I., Dobbs, T. D., & Whitaker, I. S. (2018). The association of affective disorders and facial scarring: Systematic review and meta-analysis. *Journal of Affective Disorders*, *239*, 1–10.

Golinko, M. S., Arslanian, B., & Williams, J. K. (2017). Characteristics of 1616 consecutive dog bite injuries at a single institution. *Clinical Pediatrics*, *56*, 316–325.

Guy, N. C., Luescher, U. A., Dohoo, S. E., Spangler, E., Miller, J. B., Dohoo, I. R., & Bate, L. A. (2001). A case series of biting dogs: Characteristics of the dogs, their behaviour, and their victims. *Applied Animal Behaviour Science*, *74*, 43–57.

Hollin, C. R. (2016). *The psychology of interpersonal violence*. Chichester, West Sussex: John Wiley & Sons.

Hon, K. L. E., & Leung, A. K. (2010). Childhood accidents: Injuries and poisoning. *Advances in Pediatrics*, *57*, 33–62.

Horisberger, U., Stärk, K. D. C., Rüfenacht, J., Pillonel, C., & Steiger, A. (2004). The epidemiology of dog bite injuries in Switzerland – characteristics of victims, biting dogs and circumstances. *Anthrozoös*, *17*, 320–339.

Hull, A. M., Lowe, T., Devlin, M., Finlay, P., Koppel, D., & Stewart, A. M. (2003). Psychological consequences of maxillofacial trauma: A preliminary study. *British Journal of Oral and Maxillofacial Surgery*, *41*, 317–322.

Hsu, Y., & Serpell, J. A. (2003). Development and validation of a questionnaire for measuring behavior and temperament traits in pet dogs. *Journal of the American Veterinary Medical Association*, *223*, 1293–1300.

Hsu, Y., & Sun, L. (2010). Factors associated with aggressive responses in pet dogs. *Applied Animal Behaviour Science*, *123*, 108–123.

Jagoe, A., & Serpell, J. (1996). Owner characteristics and interactions and the prevalence of canine behaviour problems. *Applied Animal Behaviour Science*, *47*, 31–42l.

Kassiri, H., Kassiri, A., Lotfi, M., Shahkarami, B., & Hosseini, S. S. (2014). Animal bite incidence in the County of Shush, Iran. *Journal of Acute Disease*, *3*, 26–30.

King, T., Marston, L. C., & Bennett, P. C. (2012). Breeding dogs for beauty and behaviour: Why scientists need to do more to develop valid and reliable behaviour assessments for dogs kept as companions. *Applied Animal Behaviour Science*, *137*, 1–12.

Klausz, B., Kis, A., Persa, E., Miklósi, A., & Gácsi, M. (2014). A quick assessment tool for human-directed aggression in pet dogs. *Aggressive Behavior*, *40*, 178–188.

Lackmann, G. M., Draf, W., Isselstein, G., & Töllner, U. (1992). Surgical treatment of facial dog bite injuries in children. *Journal of Cranio-Maxillofacial Surgery*, *20*, 81–86.

Langley, R., Mack, K., Haileyesus, T., Proescholdbell, S., & Annest, J. L. (2014). National estimates of noncanine bite and sting injuries treated in US hospital emergency departments, 2001–2010. *Wilderness & Environmental Medicine*, *25*, 14–23.

Loder, R. T. (2019). The demographics of dog bites in the United States. *Heliyon*, *5*(3), e01360.

Lofgren, S. E., Wiener, P., Blott, S. C., Sanchez-Molano, E., Woolliams, J. A., Clements, D. N., & Haskell, M. J. (2014). Management and personality in Labrador Retriever dogs. *Applied Animal Behaviour Science*, *156*, 44–53.

Macgregor, F. C. (1990). Facial disfigurement: Problems and management of social interaction and implications for mental health. *Aesthetic Plastic Surgery*, *14*, 249–257.

McGreevy, P. D., Henshall, C., Starling, M. J., McLean, A. N., & Boakes, R. A. (2014). The importance of safety signals in animal handling and training. *Journal of Veterinary Behavior*, *9*, 382–387.

Mehrkam, L. R., & Wynne, C. D. L. (2014). Behavioral differences among breeds of domestic dogs (Canislupus familiaris): Current status of the science. *Applied Animal Behaviour Science*, *155*, 12–27.

Messam, L. L. M., Kass, P. H., Chome, B. B., & Hart, L. A. (2012). Risk factors for dog bites occurring during and outside of play: Are they different? *Preventive Veterinary Medicine*, *107*, 110–120.

Messam, L. L. M., Kass, P. H., Chome, B. B., & Hart, L. A. (2018). Factors associated with bites to a child from a dog living in the same home: A bi-national comparison. *Frontiers in Veterinary Science*, *5*, Article 66.

Moxon, R., Whiteside, H., & England, G. C. (2016). Incidence and impact of dog attacks on guide dogs in the UK: An update. *The Veterinary Record*, *178*, 367.

Murphy, C. G., McGuire, C. M., O'Malley, N., & Harrington, P. (2010). Cow-related trauma: A 10-year review of injuries admitted to a single institution. *Injury, International Journal of the Care of the Injured*, *41*, 548–550.

Náhlík, J., Baranyiová, E., & Tyrlík, M. (2011). Dog bites to children in the Czech Republic: The risk situations. *Acta Veterinaria Brno*, *79*, 627–636.

Nielsen, C. S., Staud, R., & Price, D. D. (2009). Individual differences in pain sensitivity: Measurement, causation, and consequences. *The Journal of Pain*, *10*, 231–237.

O'Brien, D. C., Andre, T. B., Robinson, A. D., Squires, L. D., & Tollefson, T. T. (2015). Dog bites of the head and neck: An evaluation of a common pediatric trauma and associated treatment. *American Journal of Otolaryngology – Head and Neck Medicine and Surgery*, *36*, 32–38.

Ó Súilleabháin, P. (2015). Human hospitalisations due to dog bites in Ireland (1998–2013): Implications for current breed specific legislation. *Veterinary Journal*, *204*, 357–359.

O'Sullivan, E., Jones, B., O'Sullivan, K., & Hanlon, A. J. (2008). The management and behavioural history of 100 dogs reported for biting a person. *Applied Animal Behaviour Science*, *114*, 149–158.

Omoke, N. I., & Onyemaechi, N. O. C. (2018). Incidence and pattern of dog bite injuries treated in the emergency room of a teaching hospital South East Nigeria. *African Journal of Medical and Health Sciences*, *17*, 35–40.

Ovais, H., Adil, H., Darzi, M. A., & Beenish, M. (2017). Analysis of dog bite injuries in Kashmir. *Pulsus. Journal of Surgical Research*, *1*, 2–9.

Owczarczak-Garstecka, S. C., Watkins, F., Christley, R., & Westgarth, C. (2018). Online videos indicate human and dog behaviour preceding dog bites and the context in which bites occur. *Scientific Reports, 8* Article Number 7147.

Oxley, J. A., Christley, R., & Westgarth, C. (2018). Contexts and consequences of dog bite incidents. *Journal of Veterinary Behavior, 23*, 33–39.

Oxley, J. A., Christley, R., & Westgarth, C. (2019). What is a dog bite? Perceptions of UK dog bite victims. *Journal of Veterinary Behavior, 29*, 40–44.

Palacio, J., León-Artozqui, M., Pastor-Villalba, E., Carrera-Martín, F., & García-Belenguer, S. (2007). Incidence of and risk factors for cat bites: A first step in prevention and treatment of feline aggression. *Journal of Feline Medicine and Surgery, 9*, 188–195.

Park, J. W., Kim, D. K., Jung, J. Y., Lee, S. U., Chang, I., Kwak, Y. H., & Hwang, S. (2019). Dog-bite injuries in Korea and risk factors for significant dog-bite injuries: A 6-year cross-sectional study. *PLoS One, 14*, e0210541.

Peters, V., Sottiaux, M., Appelboom, J., & Kahn, A. (2004). Posttraumatic stress disorder after dog bites in children. *Journal of Pediatrics, 144*, 121–122.

Pirrone, F., Pierantoni, L., Mazzola, S. M., Vigo, D., & Albertini, M. (2015). Owner and animal factors predict the incidence of, and owner reaction toward, problematic behaviors in companion dogs. *Journal of Veterinary Behavior: Clinical Applications and Research, 10*, 295–301.

Podberscek, A. L., & Serpell, J. A. (1996). The English Cocker Spaniel: Preliminary findings on aggressive behaviour. *Applied Animal Behaviour Science, 47*, 75–89.

Podberscek, A. L., & Serpell, J. A. (1997). Environmental influences on the expression of aggressive behaviour in English Cocker Spaniels. *Applied Animal Behaviour Science, 52*, 215–227.

Polo, G., Calderón, N., Clothier, S., & de Casssia Maria Garcia, R. (2015). Understanding dog aggression: Epidemiologic aspects. In memoriam, Rudy de Meester (1953–2012). *Journal of Veterinary Behavior, 10*, 525–534.

Ponsich, A., Goutard, F., Sorn, S., & Tarantola, A. (2016). A prospective study on the incidence of dog bites and management in a rural Cambodian, rabies-endemic setting. *Acta Tropica, 160*, 62–67.

Raghavan, M. (2008). Fatal dog attacks in Canada, 1990–2007. *Canadian Veterinary Journal, 49*, 577–581.

Rajshekar, M., Blizzard, L., Julian, R., Williams, A., Tennant, M., Forrest, A., Walsh, L. J., & Wilson, G. (2017). The incidence of public sector hospitalisations due to dog bites in Australia 2001–2013. *Australian and New Zealand Journal of Public Health, 41*, 377–380.

Ramos, D. (2019). Common feline problem behaviors: Aggression in multi-cat households. *Journal of Feline Medicine and Surgery, 21*, 221–233.

Rezac, P., Rezac, K., & Slama, P. (2015). Human behavior preceding dog bites to the face. *Veterinary Journal, 206*, 284–288.

Roll, A., & Unshelm, J. (1997). Aggressive conflicts amongst dogs and factors affecting them. *Applied Animal Behaviour Science, 52*, 229–242.

Rosado, B., García-Belenguer, S., León, M., & Palacio, J. (2009). A comprehensive study of dog bites in Spain, 1995–2004. *Veterinary Journal, 179*, 383–391.

Rumsey, N., & Harcourt, D. (2004). Body image and disfigurement: Issues and interventions. *Body Image, 1*, 83–97.

Saadi, R., Oberman, B. S., & Lighthall, J. G. (2018). Dog-bite–related craniofacial fractures among pediatric patients: A case series and review of the literature. *Craniomaxillofacial Trauma & Reconstruction, 11*, 249–255.

Sarcey, G., Ricard, C., Thelot, B., & Beata, C. (2017). Descriptive study of dog bites in France: Severity factors, factors of onset of sequelae, and circumstances. Results of a survey conducted by InVS and Zoopsy in 2009–2010. *Journal of Veterinary Behavior, 22*, 66–74.

Shen, J., Li, S., Xiang, H., Lub, S., & Schwebel, D. C. (2014). Antecedents and consequences of pediatric dog-bite injuries and their developmental trends: 101 cases in rural China. *Accident Analysis and Prevention, 63*, 22–29.

Shepherd, K. (2009). Ladder of aggression. In D. F. Horwitz & D. S. Mills (Eds.), *BSAVA Manual of Canine and Feline Behavioural Medicine*. Quedgeley, UK: British Small Animal Veterinary Association, pp. 13–16.

Svartberg, K. (2006). Breed-typical behaviour in dogs—historical remnants or recent constructs? *Applied Animal Behaviour Science*, 96, 293–313.

Takeuchi, Y., & Mori, Y. (2006). A comparison of the behavioral profiles of purebred dogs in Japan to profiles of those in the United States and the United Kingdom. *Journal of Veterinary Medical Science*, 68, 789–796.

Takeuchi, Y., & Mori, Y. (2009). Behavioral profiles of feline breeds in Japan. *Journal of Veterinary Medical Science*, 71, 1053–1057.

Tonoike, A., Nagasawa, M., Mogi, K., Serpell, J. A., Ohtsuki, H., & Kikusui, T. (2015). Comparison of owner-reported behavioral characteristics among genetically clustered breeds of dog (Canis familiaris). *Scientific Reports*, 5, Article Number 17710.

van Delft, E. A. K., Thomassen, I., Schreuder, A. M. M., & Sosef, M. L. (2019). The dangers of pets and horses, animal related injuries in the Emergency department. *Trauma Case Reports*, 20, Article 100179.

van der Borg, J. A. M., Beerda, B., Ooms, M., de Souza, A. S., van Hagen, M., & Kemp, B. (2010). Evaluation of behaviour testing for human directed aggression in dogs. *Applied Animal Behaviour Science*, 128, 78–90.

van den Berg, S. M., Heuven, H. C. M., van den Berg, L., Duffy, D. L., & Serpell, J. A. (2010). Evaluation of the C-BARQ as a measure of stranger-directed aggression in three common dog breeds. *Applied Animal Behaviour Science*, 124, 136–141.

Vučinić, M., & Vučićević, M. (2019). Children are victims of dog bites due to irresponsible dog ownership, parenthood, and managers of school institutions in Serbia. *Journal of Veterinary Behavior*, 30, 61–68.

Wake, A., Minot, E., Stafford, K., & Perry, P. (2009). A survey of adult victims of dog bites in New Zealand. *New Zealand Veterinary Journal*, 57, 364–369.

Wani, M. M., & Sabah, Z-U. (2018). Epidemiological analysis of 1637 cases of animal bite injuries. *International Surgery Journal*, 5, 2799–2802.

West, C., & Rouen, C. (2019). Incidence and characteristics of dog bites in three remote Indigenous communities in Far North Queensland, Australia, 2006–2011. *Journal of Veterinary Behavior*, 31, 17–21.

Westgarth, C., Brooke, M., & Christley, R. M. (2018). How many people have been bitten by dogs? A cross-sectional survey of prevalence, incidence and factors associated with dog bites in a UK community. *Journal of Epidemiology and Community Health*, 72, 331–336.

Widom, C. S. (1989). Does violence beget violence? A critical examination of the literature. *Psychological Bulletin*, 106, 3–28.

Wilsson, E. (2016). Nature and nurtured – How different conditions affect the behavior of dogs. *Journal of Veterinary Behavior*, 16, 45–52.

Wilsson, E., & Sinn, D. L. (2012). Are there differences between behavioral measurement methods? A comparison of the predictive validity of two ratings methods in a working dog program. *Applied Animal Behaviour Science*, 141, 158–172.

Wrubel, K. M., Moon-Fanelli, A. A., Maranda, L. S., & Dodman, N. H. (2011). Interdog household aggression: 38 cases (2006–2007). *Journal of the American Veterinary Medical Association*, 238, 731–740.

Yeh, C. C., Liao, C. C., Muo, C. H., Chang, S. N., Hsieh, C. H., Chen, F. N., & Sung, F. C. (2012). Mental disorder as a risk factor for dog bites and post-bite cellulitis. *Injury, International Journal of the Care of the Injured*, 43, 1903–1907.

4 Pet problems
Anxiety

The state of anxiety – encompassing a myriad of terms such as fear, panic, and stress – has evolved to assist the survival of a species. Anxiety serves the purpose of alerting the individual to potential danger and prepares it to respond to the threat by flight or fright (Bateson, Brilot, & Nettle, 2011). The individual's experience of anxiety can become problematic when it becomes so frequent and protracted that it is distressing and interferes with their everyday functioning (e.g., Remes, Brayne, Van Der Linde, & Lafortune, 2016). Anxiety disorders of various types, such as phobia and obsessive-compulsive disorder (Bateson et al., 2011), are widespread in humans (e.g., Somers, Goldner, Waraich, & Hsu, 2006).

Talegón and Delgado (2011) note that for dogs, as with other animals including humans, some are more generally anxious than others. A companion animal may develop an anxiety disorder: the most common anxiety-related conditions are separation anxiety, phobias, and obsessive-compulsive disorders. As with humans, anxiety disorders can be co-morbid in animals such that, say, a dog becomes anxious when left alone and when it hears very loud noises (Tiira, Sulkama, & Lohi, 2016). An anxious dog is not only unhappy and a cause of concern for the owner, but prolonged anxiety can also bring about health problems and may ultimately shorten the animal's lifespan (Dreschel, 2010).

As they cannot tell us directly how they feel the assessment of anxiety in animals necessarily relies on the interpretation of their behaviour (Temesia, Turcsána, & Miklósia, 2014). Thus, for dogs, the repetition of behaviours such as aggression, agitated pacing, barking, destructive behaviour, panting, salivation, trembling, and even urination and defecation may be an indication of anxiety. In cats, anxiety may cause antagonism towards other cats, aggression towards humans and urine spraying (Heath, 2018). In some instances, understanding the anxiety may be relatively straightforward: for example, if the animal is distressed when there's the rumble of thunder, then thunderstorm phobia is a safe opinion. At other times, the stimulus for the anxious behaviour is less obvious and inferences must be made.

The selective breeding of companion animals for their appearance can lead to an exaggeration of behavioural as well as physical traits. Overall, Hamilton, and Chang (2006) make the point that alongside the physical consequences of selective breeding, such as hip dysplasia in some dogs and polycystic kidney disease in certain breeds of cat, there are inherited behavioural traits such as fearfulness. Thus, some breeds of dogs, including the Basset Hound, German Shepherd, and Cocker Spaniel, are more likely to be anxious than other breeds. On the positive side, the website "Dogs and Barks" (www.dogsandbarks.com) suggests that "Dogs that usually are the most anxious, are also the most intelligent and most high-energy dogs."

Separation anxiety

When animals are separated from their natural companions, they can become highly anxious. Schwartz (2003) includes birds, cats, cattle, cetaceans, dogs, horses, goats, pigs, primates, (including humans) and sheep among the species in which separation anxiety has been observed. This form of anxiety is one of the most commonly found conditions in dogs (Overall, Dunham, & Frank, 2001) and is familiar in cats (Schwartz, 2002). The scale of the problem for dogs is considerable: Overall et al. (2001) state that "Separation anxiety is one of the most common and devastating behavioral conditions in pet dogs. It has been estimated, for instance, that at least 14% of dogs examined at typical veterinary practices in the United States have signs of separation anxiety" (p. 467). While there is some debate about an exact definition, Ogata (2016) offers a pragmatic view of canine separation anxiety as: "An anxiety-related disorder in dogs whose signs are only observed in the owner's absence or perceived absence" (p. 29).

It is clear that some animals become anxious when left alone at home. Rehn and Keeling (2011) made video recordings of dogs over three time periods (30 minutes, 2 and 4 hours) beginning 10 minutes before the owner left the house and continuing until 10 minutes after the owner's return. There were no differences in the dogs' behaviour according to the length of time they were left before the owner came home. However, there were differences in the reunion with the owner: the dogs left for 2 and 4 hours were more physically active and attentive to their returning owner as compared to the 30-minute period. It appears therefore that it is a natural state of affairs for a dog's anxiety level to correlate with the length of time it is left alone at home.

In separation anxiety disorder, the dog's natural levels of anxiety at being left alone are exaggerated by an extreme attachment to the owner. Separation anxiety disorder is manifest in the dog's following their owner about the house, constant attempts to maintain physical contact, becoming visibly distressed both as their distance from the owner increases and when the owner prepares to leave the house, and an excessive greeting when the owner returns.

Palestrini, Minero, Cannas, Rossi, and Frank (2010) made video recordings of 23 dogs, aged between 5 months and 13 years, known to have separation-related problems when left alone at home. The dogs were filmed for 20–60 minutes after the owner left. The dogs spent almost one-quarter of the time alone barking, howling, or whining; other prominent behaviours included sniffing and visual inspection of the environment, panting, and destruction. These distressed behaviours began within 10 minutes of the owner leaving: some behaviours, such as barking and sniffing tended to decrease over time, while panting increased over time. Palestrini et al. suggest that three separate states – discomfort, fear, anxiety – could account for the separation issues and therefore indicting that different clinical syndromes may have been incorrectly been grouped under generic heading of anxiety. If correct, this view has implications for the assessment and treatment of canine separation anxiety.

A Norwegian study by Storengen, Boge, Strøm, Løberg, and Lingaas (2014) presents a description of 215 dogs diagnosed with separation anxiety disorder. As in the Palestrini et al. study, the most common behaviours were excessive barking, seen in 163 dogs, followed by destructive behaviour in 71 dogs. Storengen et al. found that the other indicative behaviours included hyperactivity and inappropriate urination and defecation. Storengen et al. also note that over one-half of the dogs slept in the owner's bed at night; as noted in Chapter 3, a feature of other canine problems such as aggression.

As in many judgements about behaviour, a difficulty lies in defining exactly what constitutes "excessive." Thus, Flint, Minot, Stevenson, Perry, and Stafford (2013) took an empirical approach to the issue of excessive barking. They recorded, over a 5-day period, the barking of 40 dogs left alone at home for at least 8 hours a day. They found that in the 8-hour period the dogs barked an average of between four and five times, with a mean duration of 30 seconds. The average total time spent barking was 129 seconds in the 8 hours. The dogs under 5 years of age tended to bark more often than the older dogs. Flint et al. suggest that these levels of barking set a baseline against which complaints of excessive barking can be assessed.

Cannas, Frank, Minero, Godbout, and Palestrini (2010) also adopted an empirical approach in looking at anxiety in puppies. When a puppy leaves the litter, it is normal for it to display signs of separation and begin the process of becoming attached to its new owner. It is when the puppy's reactions to separation endure over the longer term that matters become problematic. Over a 2-month period, Cannas et al. made three videotape recordings of 32 puppies, aged from 50 to 118 days, when they were left alone at home. They found that three of the puppies showed signs of anxiety but that these behaviours decreased over time. Over the duration of the study, one puppy displayed protracted signs of distress when home alone: this puppy's behaviour may indicate the likelihood of future difficulties.

In a study investigating the background risk factors for separation anxiety, van Rooy, Thomson, McGreevy, and Wade (2018) used owner's responses to the C-BARQ to look at the behaviour of 226 Golden Retrievers and 247 Labrador Retrievers aged between 10 months and 15 years. They identified significant associations between separation-related behaviour and various traits including attention-seeking and excitability. The strong predictors of separation anxiety were when the dog was acquired from a pet shop or shelter rather than a service organisation (such as for guide dogs) or a breeder; age at adoption, in that dogs acquired when older than 6 years were most likely to exhibit signs of anxiety when compared to dogs who were acquired when 2 to 3 months old; finally, anxiety was more likely to be seen in castrated rather than intact males.

The stereotype of domestic cats is that, unlike dogs, they are asocial at best or antisocial at worst. However, the evidence indicates that cats form social bonds, with both their owners and with other cats (Bradshaw, 2016; Voith and Borchelt, 1986), so that they may react to separation in a way not too dissimilar to dogs. In a study of 136 cats with separation anxiety disorder, Schwartz (2002) reported a range of behavioural issues which involved excessive vocalisation, destruction of property, tail-chewing, and inappropriate urination and defecation. Schwartz records that three-quarters of the cases of inappropriate urination took place solely on the owner's bed.

Explaining separation anxiety disorder

Following a review of the literature, Ogata (2016) gives two broad explanations for canine separation anxiety disorder: (1) hyper- or over-attachment and (2) underlying fear and anxiety. From the first explanation, the separation anxiety is a consequence of the high level of attachment to the owner which started when the dog was a puppy and strengthened as it grew to an adult dog. The attachment becomes so strong that the dog continually seeks close proximity to the owner and is distressed in their absence. However, Ogata concludes that support for this explanation is questionable given the lack of definitive proof that hyper-attachment exists and that an alternative explanation may lie in "a manifestation of other underlying motivations" (p. 32).

If separation anxiety disorder is a manifestation of an underlying anxiety trait, as per the second explanation, then it may be predicted that dogs with separation anxiety disorder would display other forms of anxiety. The evidence supports this view indicating a high level of co-morbidity between separation anxiety and other anxiety states such as phobias (e.g., Overall et al., 2001). The perceived attachment problem may in fact be a learned behaviour where the animal seeks contact with the owner for comfort when an anxiety evoking situation occurs. This owner-seeking behaviour is positively reinforced over time as the owner responds to their anxious pet, eventually leading to the mistaken description of the repetitive learned behaviour as a form of attachment. This learning theory perspective is in keeping with the view, discussed in Chapter 2, that the extension of the human concept of attachment to animals is problematical (Crawford, Worsham, & Swinehart, 2006; Potter & Mills, 2015).

Phobias

In humans, a phobia is a deep fear, typically manifest as a panic attack, triggered by the presence of a specific situation or entity. Common phobias include fear of heights (*acrophobia*), enclosed spaces (*claustrophobia*), and flying (*aerophobia*). There are also phobias associated with everyday animals such as cats (*ailurophobia*), dogs (*cynophobia*), and spiders (*arachnophobia*); with the less commonly encountered animals such as snakes (*ophidiophobia*) and ladybirds (*coccinellidaephobia*); and with the more exotic such as elephants (*pachydermophobia*) and bears (*arkoudaphobia*).

In cats and dogs, a fear of loud noises is relatively widespread. Tiira et al. (2016) reported that in a survey of 3,284 dog owners, 1,287 (39.2%) said that their dog reacted fearfully to loud noises. Of these 1,287 respondents with a noise-sensitive dog, 407 said they first noticed their dog's fearful reactions to loud noises at ages ranging from 8 weeks to 10 years with a median age of 2 years. Blackwell, Bradshaw, and Casey (2013) conducted a postal survey of 3,897 dog owners in the UK asking about their dog's fear or anxiety on exposure to noise. In all, one-quarter of the owners reported that their dog showed a fear of noises. The noises most noted as eliciting a fear reaction included, from most to least frequently stated, fireworks, thunderstorms, gunshots, cars backfiring, and loud noises on television. Blackwell, Bradshaw, and Casey note that the dog's reaction is often referred to as a "noise phobia." They found that less than one-third of the owners had looked for assistance in resolving their pet's problem with noise.

While sensitivity to noise is a common issue it is not a universal trait among dogs. A Norwegian study by Storengen and Lingaas (2015) looked at sensitivity to noise (fireworks, gunshots, heavy traffic, and thunderstorms) across 5,248 dogs encompassing 17 breeds. In close agreement with the UK study by Blackwell, Bradshaw, and Casey, they found that almost 23% of the owners reported that their dog showed a fear of noise. The greatest level of fear was for fireworks, followed in decreasing order by loud noises/gunshots, thunderstorms, and heavy traffic; a fear of all three types of noise frequently co-occurred. There was a significant difference across breeds: the highest level of sensitivity was seen in the Norwegian Buhund, Irish Soft Coated Wheaten Terrier, and Lagotto Romagnolowere; the lowest was evident the Boxer, Chinese Crested, and Great Dane. The older dogs were more likely to be fearful, as were the female dogs, while neutered dogs were more likely to be fearful than intact dogs. Those dogs most fearful of noise were also more likely to show showing separation-related anxiety and fear of new situations. Given the likelihood of thunderstorms and the frequency of fireworks displays, the effects of these two types of noise on companion animals have been considered in some detail.

Thunderstorms

A fear of thunderstorms, termed *astraphobia*, is to be found across many species, including humans. Alongside companion animals, animals such as cattle and horses may react fearfully when a thunderstorm occurs and seek shelter in the interest of self-preservation. It is not unknown for animals to be struck by lightening while the high winds that often accompany storms may carry debris that has the potential to cause physical harm.

The effect of a thunderstorm on some dogs is illustrated in a case study presented by Radosta (2016). The dog in question, called Milo, is a 6-year-old, neutered, mixed Chihuahua-dachshund, a crossbreed known as a *Chiweenie*. From an early age, Milo showed signs of anxiety, such as hiding and trembling, during thunderstorms. As Milo grew older, the anxiety worsened; if the owners were at home during the storm, Milo hid, hypersalivated, trembled, and whined; when home alone, he also urinated and defecated (there were no indications of fear when he was left alone and there were no storms). Milo did not show signs of fear when there were fireworks displays. Milo's case will be returned to in the following chapter.

Fireworks

There are obvious similarities between thunderstorms and fireworks and Blackwell et al. (2013) reported that more dogs were said by owners to show fear of fireworks than to any other loud noise. A survey carried out in New Zealand by Dale, Walker, Farnworth, Morrissey, and Waran (2010) with owners of several thousand cats and dogs found that close to one-half of the animals displayed a discernible fear of fireworks. The detection of a fear response is not always straightforward so that some owners may not recognise their pet's fear. Thus, the true prevalence of some fears is liable to be even greater than the figures suggest. In addition, Bolster (2012) makes the point that it is not just cats and dogs that are afraid of fireworks. Bolster suggests that in considering the effect of fireworks "Pets, such as rabbits, guinea pigs and other 'small furries' are overlooked" (p. 387).

Stress

In the clinical literature the distinction is drawn between *acute stress* and *chronic stress*. Acute stress may be thought of as a normal physical and psychological reaction to a highly aversive event, such as an accident, a bereavement, or a threat to the person. The acute stress reaction takes place immediately or within hours of the event. The individual's initial state of shock and disorientation is followed either by withdrawal from the situation or by agitation and symptoms of panic such as tachycardia and perspiration. In some cases, partial or complete amnesia for the episode may occur. The acute stress reaction subsides and expires within hours or days of the stressful event. In contrast, a chronic stress reaction may result when the acute stress does not dissipate or if the cause of the stress remains present. The unmitigated combination of acute and chronic stress may culminate in post-traumatic stress disorder (Bryant et al., 2017).

Beerda, Schilder, van Hooff, and de Vries (1997) applied the concepts of acute and chronic stress to the welfare of canines. They make the point that a range of environmental conditions, including poor housing, harsh training, and unpredictable aversive events – as with the dogs in Seligman's experiments discussed in Chapter 1 – create stressful conditions for the dog. These environmental stressors can act to influence adversely the dog's behaviour (Beerda, Schilder, Van Hooff, De Vries, & Mol, 1999), its physiological functioning (Beerda et al., 1999), and its health (Mills, Karagiannis, & Zulch, 2014).

The domestic cat may also experience stress when placed in an unpredictable and aversive environment. Amat, Camps, and Manteca (2015) note that the effects of stress may be seen in behaviours such as aggression, compulsive behaviours as with over-grooming, urine marking, and reduced feeding to the point of stress-related anorexia and consequent medical complications. It is important to make the point, germane to many problem behaviours in both animals and humans, that anxiety is not necessarily the cause of the aberrant behaviour. In cats, for example, urinary house soiling may be caused by territorial disputes with other cats or disease.

Overall and Dunham (2002) suggest that repetitive behaviour may be seen as an indicator of a cat or dog's obsessive-compulsive disorder. As with all animals, including humans, where repetitive behaviour is displayed, the case for obsessive-compulsive disorder lies in the disruptive effect of the repetitive behaviour on normal, everyday functioning. However, the case is not fully made that repetitive behaviour is equivalent to obsessive-compulsive disorder as would be diagnosed in humans. Frank (2013) argues that: "Data on compulsive disorders in dogs and cats are scarce and incomplete," (p. 130) which means that any attempt to equate what is known about compulsive disorders in animals and humans falls short of the necessary scientific evidence.

Frank points to the problem that if the term *obsessive-compulsive* is used in veterinary medicine synonymously with repetitive behaviour then there is a risk that findings from animal research are erroneously used to inform models for obsessive-compulsive disorder in humans (see Dodman & Shuster, 2005). In support of this Frank's position, Tynes and Sinn (2014) caution that stereotypic behaviours have not been shown to be the same as obsessive-compulsive behaviours given that too little is understood about their respective aetiologies.

The importance of understanding and ameliorating stress is highlighted by Hekman, Karas, and Sharp (2014), who make that point that some animals receiving treatment in a veterinary hospital may well find the experience stressful. A period in hospital precipitates stressors such separation from the owner, disruption of regular routines, and exposure to novel and intrusive surroundings. In order to make an accurate diagnosis and deliver appropriate treatment, the veterinary practitioner may need to disentangle the effects of stress from the pathological symptoms that led to hospitalisation.

In this and the previous chapter the two principal classes of behaviour problems in companion animals have been considered. There are other problems – such as dogs jumping, digging, and pulling on the lead, while cats may have litter tray issues and destructive scratching – but with all problems the net result is the same: an unhappy owner and an animal that may be seen as a chore at best or disposable at worst. One of the basic premises of applied psychology is that behaviour can be changed, so how does this maxim apply to solving the problems posed by companion animals?

References

Amat, M., Camps, T., & Manteca, X. (2015). Stress in owned cats: Behavioural changes and welfare implications. *Journal of Feline Medicine and Surgery, 18*, 577–586.

Bateson, M., Brilot, B., & Nettle, D. (2011). Anxiety: An evolutionary approach. *Canadian Journal of Psychiatry, 56*, 707–715.

Beerda, B., Schilder, M. B. J., van Hooff, J. A. R. A. M., de Vries, H. W., & Mol, J. A. (1999). Chronic stress in dogs subjected to social and spatial restriction. I. behavioral responses. *Physiology & Behavior, 66*, 233–242.

Beerda, B., Schilder, M. B. J., Bernadina, W., van Hooff, J. A. R. A. M., de Vries, H. W., & Mol, J. A. (1999). Chronic stress in dogs subjected to social and spatial restriction. II. Hormonal and immunological responses. *Physiology & Behavior, 66*, 243–254.

Beerda, B., Schilder, M. B. J., van Hooff, J. A. R. A. M., & de Vries, H. W. (1997). Manifestations of chronic and acute stress in dogs. *Applied Animal Behaviour Science, 52*, 307–319.

Blackwell, E. J., Bradshaw, J. W. S., & Casey, R. A. (2013). Fear responses to noises in domestic dogs: Prevalence, risk factors and co-occurrence with other fear related behaviour. *Applied Animal Behaviour Science, 145*, 15–25.

Bolster, C. (2012). Fireworks are no fun for pets. *Veterinary Nursing Journal, 27*, 387–390.

Bradshaw, J. W. S. (2016). Sociality in cats: A comparative review. *Journal of Veterinary Behavior, 11*, 113–124.

Bryant, R. A., Creamer, M., O'Donnell, M., Forbes, D., McFarlane, A. C., Silove, D., & Hadzi-Pavlovic, D. (2017). Acute and chronic posttraumatic stress symptoms in the emergence of posttraumatic stress disorder: A network analysis. *Journal of the American Medical Association: Psychiatry, 74*, 135–142.

Cannas, S., Frank, D., Minero, M., Godbout, M., & Palestrini, C. (2010). Puppy behavior when left home alone: Changes during the first few months after adoption. *Journal of Veterinary Behavior, 5*, 94–100.

Crawford, E. K., Worsham, N. L., & Swinehart, E. R. (2006). Benefits derived from companion animals, and the use of the term "attachment". *Anthrozoös, 19*, 98–112.

Dale, A. R., Walker, J. K., Farnworth, M. J., Morrissey, S. V., & Waran, N. K. (2010). A survey of owners' perceptions of fear of fireworks in a sample of dogs and cats in New Zealand. *New Zealand Veterinary Journal, 58*, 286–291.

Dodman, N. H., & Shuster, L. (2005). Animal models of obsessive-compulsive behavior: A neurobiological and ethological perspective. In J. S. Abramowitz & A. C. Houts (Eds.), *Concepts and controversies in obsessive-compulsive disorder* (pp. 53–71). Boston, MA: Springer.

Dreschel, N. A. (2010). The effects of fear and anxiety on health and lifespan in pet dogs. *Applied Animal Behaviour Science, 125*, 157–162.

Flint, E. L., Minot, E. O., Stevenson, M., Perry, P. E., & Stafford, K. J. (2013). Barking in home alone suburban dogs (*Canis familiaris*) in New Zealand. *Journal of Veterinary Behavior, 8*, 302–305.

Frank, D. (2013). Repetitive behaviors in cats and dogs: Are they really a sign of obsessive-compulsive disorders (OCD)? *Canadian Veterinary Journal, 54*, 129–131.

Heath, S. (2018). Understanding feline emotions... and their role in problem behaviours. *Journal of Feline Medicine and Surgery, 20*, 437–444.

Hekman, J. P., Karas, A. Z., & Sharp, C. R. (2014). Psychogenic stress in hospitalized dogs: Cross species comparisons, implications for health care, and the challenges of evaluation. *Animals: An Open Access Journal from MDPI, 4*, 331–347.

Mills, D., Karagiannis, C., & Zulch, H. (2014). Stress—its effects on health and behavior: A guide for practitioners. *Veterinary Clinics: Small Animal Practice, 44*, 525–541.

Ogata, N. (2016). Separation anxiety in dogs: What progress has been made in our understanding of the most common behavioral problems in dogs? *Journal of Veterinary Behavior, 16*, 28–35.

Overall, K. L., & Dunham, A. E. (2002). Clinical features and outcome in dogs and cats with obsessive-compulsive disorder: 126 cases (1989–2000). *Journal of the American Veterinary Medical Association, 221*, 1445–1452.

Overall, K. L., Dunham, A. E., & Frank, D. (2001). Frequency of nonspecific clinical signs in dogs with separation anxiety, thunderstorm phobia, and noise phobia, alone or in combination. *Journal of the American Veterinary Medical Association, 219*, 467–473.

Overall, K. L., Hamilton, S. P., & Chang, M. L. (2006). Understanding the genetic basis of canine anxiety: Phenotyping dogs for behavioral, neurochemical, and genetic assessment. *Journal of Veterinary Behavior, 1*, 124–141.

Palestrini, C., Minero, M., Cannas, S., Rossi, E., & Frank, D. (2010). Video analysis of dogs with separation-related behaviors. *Applied Animal Behaviour Science, 124*, 61–67.

Potter, A., & Mills, D. S. (2015). Domestic cats (*Felis silvestris catus*) do not show signs of secure attachment to their owners. *PLoS One, 10*, e0135109.

Radosta, L. (2016, July). Storm phobia in dogs. *Veterinary Team Brief*, 33–37.

Remes, O., Brayne, C., Van Der Linde, R., & Lafortune, L. (2016). A systematic review of reviews on the prevalence of anxiety disorders in adult populations. *Brain and Behavior, 6*, e00497.

Rehn, T., & Keeling, L. J. (2011). The effect of time left alone at home on dog welfare. *Applied Animal Behaviour Science, 129*, 129–135.

Schwartz, S. (2002). Separation anxiety syndrome in cats: 136 cases (1991–2000). *Journal of the American Veterinary Medical Association, 220*, 1526–1532.

Schwartz, S. (2003). Separation anxiety syndrome in dogs and cats. *Journal of the American Veterinary Medical Association, 222*, 1028–1033.

Somers, J. M., Goldner, E. M., Waraich, P., & Hsu, L. (2006). Prevalence and incidence studies of anxiety disorders: A systematic review of the literature. *Canadian Journal of Psychiatry, 51*, 100–113.

Storengen, L. M., Boge, S. C. K., Strøm, S. J., Løberg, G., & Lingaas, F. (2014). A descriptive study of 215 dogs diagnosed with separation anxiety. *Applied Animal Behaviour Science, 159*, 82–89.

Storengen, L. M., & Lingaas, F. (2015). Noise sensitivity in 17 dog breeds: Prevalence, breed risk and correlation with fear in other situations. *Applied Animal Behaviour Science, 171*, 152–160.

Talegón, M. I., & Delgado, B. A. (2011). Anxiety disorders in dogs. In V. V. Kalinin (Ed.), *Anxiety disorders* (pp. 261–280). Rijeka, Croatia: InTech.

Temesia, A., Turcsána, B., & Miklósia, A. (2014). Measuring fear in dogs by questionnaires: An exploratory study toward a standardized inventory. *Applied Animal Behaviour Science, 161*, 121–130.

Tiira, K., Sulkama, S., & Lohi, H. (2016). Prevalence, comorbidity, and behavioral variation in canine anxiety. *Journal of Veterinary Behavior, 16*, 36–44.

Tynes, V. V., & Sinn, L. (2014). Abnormal repetitive behaviors in dogs and cats: A guide for practitioners. *Veterinary Clinics: Small Animal Practice, 44*, 543–564.

van Rooy, D., Thomson, P. C., McGreevy, P. D., & Wade, C. M. (2018). Risk factors of separation-related behaviours in Australian retrievers. *Applied Animal Behaviour Science, 209*, 71–77.

Voith, V. L., & Borchelt, P. L. (1986). Social behavior of the domestic cat. In V. L. Voith & P. L. Borchelt (Eds.), *Readings in companion animal behavior* (pp. 248–257). Trenton, NJ: Veterinary Learning Systems.

5 Solving pet problems

In the previous chapter, the various behaviours exhibited by animal companions that we perceive as problematic were discussed. There are two ways to circumvent these problems: first, nullify the behaviour through biological intervention; second, increase the animal's socialisation through training. This chapter looks at both of these approaches, with an emphasis on training to reduce aggression and anxiety, two of the most common and distressing problems.

The unkindest cut of all

There are three strategies based on biological intervention commonly used to change an animal's behaviour: 1) breeding programmes, 2) neutering, 3) drugs.

Breeding programmes

As some dogs are known to be aggressive, it should be possible to reduce the prominence of the aggression by excluding aggressive dogs from the breeding population. van der Borg, Graat, and Beerda (2017) reported an evaluation of a Dutch selective breeding policy where Rottweilers assessed as highly fearful and aggressive were removed from the pedigree breeding pool. This programme, run under the auspices of the Dutch Kennel Club, excluded aggressive dogs from obtaining pedigree certificates of the Fédération Cynologique Internationale (FCI). They compared owners' assessments of their dogs' behaviour with a group of 395 Rottweilers with pedigree certificates and 427 "look-alike" Rottweilers with no pedigree certificate: "Dutch Rottweilers with a FCI pedigree certificate had reduced (owner-reported) fear and/or aggression towards strange people and non-social fear in comparison to look-a-likes, which we attribute to the breeding policy of the Dutch Kennel Club" (p. 85).

Neutering

It is common practice to sterilise male and female dogs – *castration* and *ovariohysterectomy*, collectively termed *gonadectomy* – both to address behaviour problems, including aggression, and to make dogs better-behaved companions. Farhoody et al. (2018) conducted a survey of 13,370 dog owners contrasting the aggressive behaviour of dogs with and without a gonadectomy towards familiar people, towards strangers and towards other dogs. The analysis showed no significant relationship between any of the three types of aggressive behaviour and either gonadectomy status or the dog's age at gonadectomy. Farhoody et al. note that their findings are in accord with a substantial body of research on the effects of gonadectomy on aggressive behaviour. The practice of surgical sterilisation through gonadectomy may have detrimental long-term effects on both the dog's health (Zwida & Kutzler, 2016) and social behaviour (Kaufmann, Forndran, Stauber, Woerner, & Gansloßer, 2017). There are advocates of alternative procedures such as a vasectomy for male dogs and a partial spay for bitches that removes the uterus and leaves the ovaries intact (Brent & Kutzler, 2018).

Drugs

There are a range of medications, such as fluoxetine (Prozac for humans), sertraline, and clomipramine, used in the treatment of aggression and other behavioural problems. Siracusa (2016) presents a case study of two companion mixed-breed female dogs, a 2-year-old and a 4-year-old, who were aggressive to each other when food was present or if one approached the other when resting. The intervention involved serotonergic medication to decrease anxiety, arousal, and impulsivity for both dogs; along with clomipramine hydrochloride, a tricyclic antidepressant, for one dog. The drug treatment was accompanied by behavioural training to enable the owners to modify the dogs' actions as necessary. At a 6-month follow-up, there were no aggressive incidents. The use of drugs and behaviour modification is a well-established combination in the treatment of behavioural disorders.

Training methods and their effectiveness

While most companion animals happily fit into our everyday lives some training, typically to increase desired behaviour and reduce or eliminate problematic behaviour, is usually required for their complete socialisation and a problem-free existence. Indeed, González-Martínez et al. (2019) suggest that puppy training plays an important part, for owner and dog, in the development of a lasting relationship.

Blackwell, Twells, Seawright, and Casey (2008) conducted a survey of the types of training preferred by 192 dog owners. They reported that 58% of the sample said they trained their dog themselves while "General obedience classes were attended by 40% of owners, and 27% attended puppy socialization classes. Agility or flyball classes were attended by 12% of owners, and 5% of dogs were taken to handling or showing classes" (p. 209). In addition to classes, there are any number of books on how to train various animals including cats (e.g., Johnson-Bennett, 2011), dogs (e.g., Whittaker, 2005), and rabbits (e.g., Isbell & Pavia, 2009).

However, it is not just a straightforward matter of recognising the need for training, the way owners make causal attributions about their dogs' behaviour is important. An owner who attributes the cause of their dog's aggression to other dogs and people is very different to an owner to see the cause as a lack of training. Wong and Cheong (2010) conducted a survey of 50 male and 50 female dog owners from Selangor and Kuala Lumpur. They found that owners who saw problem behaviour as stable were more likely to report higher levels of behaviour problems: these same owners were more likely to use ineffective methods of discipline. Thus, the owners' perceptions and attributions influence their capability to discipline their dog and in turn change the dog's behaviour.

Set against the vagaries of the owner, what methods of training are available? As shown in Table 5.1, the most commonly used training methods divide into two types:

Table 5.1 Methods of dog training based on reward and aversion

Reward	Aversion
Food treats	Bark-activated electronic collar
Stroking/petting	Jerking back on lead
Clicker training	Withholding treats
Verbal praise	Shutting away
	Bark-activated citronella collar
	Verbal punishment (shouting)
	Electric fence
	Physical punishment
	Remote activated electric collar
	Choke chain

first, methods based on rewarding appropriate behaviours; second, methods that deliver aversive stimuli or withhold rewards to reduce unwanted behaviour. These methods correspond exactly with Thorndike's Law of Effect and Skinner's principles of behaviour change discussed in Chapter 1.

Increasing appropriate behaviour with treats or praise is *positive reinforcement*, while attempting to *increase* appropriate behaviour by withholding treats is *negative reinforcement*. Aversive training methods may be used to try *reduce* inappropriate behaviour by causing pain, as with a jerk lead, which is *positive punishment*; while taking away something the animal values, such as shutting it away to deny social contact, is *negative punishment*.

Reward-based methods

Rewards-based training relies on the positive reinforcement of specific behaviours, such as sitting or walking on the lead, so that they occur in response to the trainer's verbal and nonverbal cues (Mills, 2005). The immediate issue is what to use as a reward to reinforce the behaviour most effectively. Fukuzawa and Hayashi (2013) compared three effects of three types of reward – food, stroking, and praise – on dogs' learning. The experimental design allocated 15 dogs to three different reward groups all trained to "sit" and "stay" by the same trainer using an identical method. It was found that in the early stages of training only the use of food as a reward shortened the time for completion of the dog's response to the command. The rewarding effect of food was not, however, evident in the later stages of training.

Given that different dogs have diverse food preferences (Vicars, Miguel, & Sobie, 2014), it follows that, as Riemer, Ellis, Thompson, and Burman (2018) point out, the description of a reward as "food" covers a range of possibilities. Riemer et al. considered the effects of the quality and quantity of food used as a reward on dogs' completion of a running task. While the quantity of the reward had no effect, the dogs ran significantly faster when the reward was food of a higher quality.

The Fukuzawa and Hayashi study highlights the issue of continual versus intermittent reinforcement schedules. When a continuous schedule is in operation *every* occurrence of the behaviour is rewarded. With an intermittent schedule the behaviour may be reinforced after a set time or number of behaviours (a *fixed* schedule) or after a variable time or number of behaviours (an *intermittent* schedule). The use of schedules is of practical importance in behaviour acquisition and change (e.g., Hulac, Benson, Nesmith,

& Shervey, 2016). It may be effective to use continuous reinforcement at the early stages of behaviour acquisition followed by a variable schedule to maintain the behaviour. The importance of the timing of training sessions is highlighted by Demant, Ladewig, Balsby, and Dabelsteen (2011), who compared the effects of several combinations of frequency and duration of training sessions. They found that shorter training sessions held once or twice a week are more effective than long daily sessions.

The process of learning to socialise with their group is a key part of early learning that all animals must undertake. As seen with Harlow's monkeys discussed in Chapter 1, failure in early social learning may well lead to impaired life and social skills resulting in diminished engagement with peers and the wider environment (Dietz, Arnold, Goerlich-Jansson, & Vinke, 2018). Vaterlaws-Whiteside and Hartmann (2017) devised a "new nest" training programme for puppies aged 6 weeks and younger. The programme is based on the premise that "For young puppies, socialization is facilitated by people through the introduction of various social and environmental stimuli" (pp. 55–56). The programme moves through the presentation of a range of socialisation stimuli – spanning the stages of tactile, auditory, visual, interaction with people, and interaction with the environment – designed to follow the puppy's physiological and behavioural development from birth to 6 weeks of age. The evaluation of the programme showed that it was successful and had an enduring effect as the puppies grew older. Vaterlaws-Whiteside and Hartmann suggest the programme may be particularly useful when puppies are raised in animal shelters or as working dogs, such as guide dogs, raised in kennels. This programme is notable for mimicking naturally occurring contingencies which provide rewards for the young dogs.

There is no doubt that dog training can be a time-consuming enterprise. There are several gadgets commercially available which are intended to make the task easier to perform.

Training gadgets

As an alternative to anti-barking devices (Cronin et al., 2003) advances in technology were put to use by Yin, Fernandez, Pagan, Richardson, and Snyder (2008) to reduce barking, jumping, and rushing to the door when people call. Yin et al. taught dogs to eat from a remote-control food dispenser. When a tone sounded, food was available, thereby forming association between the tone emitted by the dispenser and food. This use of association is utilisation of Pavlov's classical conditioning such that the tone becomes the conditioned stimulus for the conditioned reward. Once the association is established, it provides the basis for sequence of further training to shape the dog's behaviour so that it ran to a rug, lay down, and stayed for a minute while household door noises such as loud knocking and ringing doorbells were audible.

Protopopova, Kisten, and Wynne (2016) employed a similar strategy, based on reinforcement rather than punishment as is typical with remote devices, to reduce home-alone nuisance barking. When the dog was home alone, a computer-controlled device delivered food contingent on preset intervals of not barking. The strategy was successful in reducing barking and the effect was maintained when the intervals between the delivery of food were lengthened (i.e., variable interval reinforcement).

Clicker training

The publication of the book, *Don't Shoot the Dog!* by Karen Pryor (1999), a dolphin trainer and later a dog trainer, led to *clicker training* becoming popular in both dog and horse training, later extending to cats (Kogan, Kolus, & Schoenfeld-Tacher, 2017). A clicker is a hand-held device which when pressed makes a "click-clack" sound. The clicker is pressed by the trainer, immediately followed by a food reward, when the animal carries out a chosen behaviour (Figure 5.1).

The principles underpinning clicker training are somewhat unclear, not helped by the use of imprecise terminology (Dorey & Cox, 2018), and there are several hypotheses as it how it may function (Feng, Howell, & Bennett, 2016). Yet further, the mode of application of clicker training varies across trainers, which sets stern challenges to the evaluation of its effectiveness (Feng, Howell, & Bennett, 2018). In keeping with this air

Figure 5.1 Clicker dog training.
Source: Photograph by SpeedKingz. Shutterstock.

of uncertainly, the evaluations are indeterminate as to the effectiveness of clicker training (e.g., Chiandetti, Avella, Fongaro, & Cerri, 2016; Fugazza & Miklósi, 2015; Smith & Davis, 2008).

Aversion-based methods

Following a survey of the training methods used by 140 dog owners in Philadelphia, Herron, Shofer, and Reisner (2009) made the distinction between *confrontational* and *non-confrontational* training methods used to deal with aggression to people and dogs, barking, and separation anxiety. Non-confrontational methods such as using food treats and clicker training are based on reward; as shown in Table 5.2, the confrontational methods involve direct belligerent physical contact with the dog.

Electric shock collars

The composition and use of the electric shock collar (or e-collar), which can deliver shocks of up to 6,000 volts, is explained by Schilder and van der Borg (2004):

> The electric collar consists of a collar that includes a battery and electrodes, and a remote control, through which the trainer can deliver shocks of various durations and intensities to the dog. Some types of collar can be tuned finely to the sensitivity of the dog being trained; other types possess only a limited possibility of adaptation. Some collars include a feature to warn the dog by sounding a beep, before a shock is delivered. Shock duration may vary from 1/1000 of 1–30 s; shock intensity will vary with coat structure and humidity, but in general a current of a few thousand volts is used. (p. 320)

In addition, the shock can be bark activated so that the dog receives a shock when it barks. An electronic boundary fence collar consists of a wire, which functions as an aerial, laid around the perimeter of, say, a garden. The dog wears a minute receiver on its collar which picks up a signal from the wire when the dog gets close to it: a sound warns the dog not to approach then, if the dog continues, a shock is administered. The shock collar can be used as a punishment, although it can also be used in training when the shock is part of a negative reinforcement contingency such that the dog obeys to avoid the pain. Schilder and van der Borg conducted a study looking at the effects of the shock collar used in training 32 German Shepherds to work as guard dogs. They found that, not

Table 5.2 Examples of confrontational training methods (Herron et al., 2009)

"Alpha roll": Rolled dog onto its back and hold it down
Force down: Step on the leash to force the dog to lie down
Hit, grab, or kick dog
Knee dog in chest in mid-air to deter jumping
Neck jab: Jab the dog on neck or side with fingers
Rub the dog's nose in the spot where the house has been soiled
Remote-activated and bark-activated electric shock collar
Spray the dog with water pistol/spray bottle
Growl at dog
Forcibly expose the dog to frightening stimulus such as a loud noise

surprisingly, in the short term the shocked dogs found the training painful and stressful; in the long term the dogs continued to show signs of stress and a particular aversion to their handler. Schilder and van der Borg argue that the use this training method acts against the dog's welfare.

A survey of 1,251 dog owners in France, where there are no restrictions on the use of shock collars, was reported by Masson, Nigron, and Gaultier (2018). They found that over one-quarter of the owners had used an electronic device, including bark-activated collars (149 owners), electronic boundary fence collars (56 owners), and remote-controlled collars (178 owners). The shock collars were most evident when the dog weighed over 40 kg, was not neutered, and was used for hunting or security. The majority of owners used the collars without professional advice and before trying other solutions. Masson, Nigron, and Gaultier make the point that while their survey suggests a high level of use of shock collars, given their potentially detrimental effects on animal welfare there are no regulations to govern their usage.

Blackwell, Bolster, Richards, Loftus, and Casey (2012) conducted a similar survey in England, asking 3,897 dog owners about their use of electronic collars. They reported that 133 owners (3.3% of the sample) used remote-activated collars, 54 owners (1.4%) used bark-activated collars, and 36 (0.9%) used electronic boundary fences.

Opposition to electric devices

The European Society of Veterinary Clinical Ethology (ESVCE) is an organisation whose website states that it aims "To promote and support scientific progress in veterinary behaviour medicine and comparative clinical ethology." ESVCE published a position statement on the advantages and disadvantages of the use of electronic training devices with dogs (Masson et al., 2018). A review of the research evidence alongside consideration of the moral and ethical issues led to the conclusion that:

> Training with e-collars is associated with numerous well-documented risks concerning dog health, behavior, and welfare. When e-collars are used to treat behavior problems, there is a risk of such problems worsening and/or additional problems emerging. (p. 74)

Thus, ESVCE reach the position of calling for a European ban, applicable in all member states, on the advertising, distribution, and sale of e-collars. Animal welfare acts or similar could legally enforce such a ban, which would involve fines and custody for persistent offenders.

The Norwegian Scientific Committee for Food and Environment (*Vitenskapskomiteen for mat og miljø*; VKM) completed a review of the use of electric devices with animals (VKM, Mejdell, Basic, & Bøe, 2017), as well as electric collars and fences; the Norwegian review includes the use of electric shocks for immobilisation as used for example in the fishing industry and as an alternative to anaesthetic in some surgical procedures. This review also covers the use of electric shocks to modify horse behaviours such as crib-biting and wind sucking, in sperm collection for breeding ("electro-ejaculation"), and the use of cattle goads (i.e., cattle prods). The VKM review concludes with a summary of the application of each device used to administer shocks. The main overall issues they highlight are a lack of research evidence in some areas and varying levels of apprehension regarding the short- and long-term effects of electric devices on the animal.

The concerns about the use of electronic devices has led to a ban on the use of shock collars in Austria, some Australian states (New South Wales and Southern Australia), Denmark, Germany, Norway, Slovenia, Sweden, and Switzerland. In the UK the use of electric shock collars is prohibited in Scotland and Wales. In England, in 2018 The Department of Environment, Food and Rural Affairs (Defra) conducted a consultation exercise on the proposed ban on electronic training collars (e-collars) for cats and dogs (Department of Environment, Food and Rural Affairs, 2018). The consultation received a large number of responses both from organisations – there was support for the ban from The Kennel Club and RSPCA and some reservations expressed by The Universities Federation for Animal Welfare (UFAW) – and members of the general public. The summary document also note that the use of electronic collars is: "Proscribed by the British Veterinary Association (BVA, 2006), the European Society of Veterinary Clinical Ethology … and the Australian Veterinary Association" (p. 20).

The outcome of the consultation was a decision to amend the Animal Welfare Act 2006 so as to ban the use of hand-held remote-controlled e-collar devices in England. The ban did not extend to fencing containment systems, the use of which would be held under review.

Feline training

The cat's independent nature makes the need for training less demanding although there are situations where this independence must be compromised so that training becomes a necessity. Pratsch et al. (2018) note that when a cat is transported to the vets not only must it endure confinement in a carrier, the movement from home to the vets will involve strange smells and noises then handling by strangers in an unfamiliar environment. The physiological state of the cat after this ordeal is not conducive to veterinary examination and may even disguise certain conditions. Pratsch et al. devised and evaluated a reward-based training programme in which the cats progressed through seven stages from approaching the carrier, staying in the carrier, and finally being transported in the carrier. The programme was successful in that the cats showed fewer signs of stress and increased compliance during the veterinary examination. There are similar approaches to desensitising cats to the procedures involved in taking blood samples (Lockhart, Wilson, & Lanman, 2013).

In a quite different context, animals used in laboratory research may find the routines of the laboratory environment stressful to the point that it adversely affects the quality, and hence worth, of the data (Bailey, 2018). The need to contain a cat in order to transport it from research location to location can create a situation which the cat may find stressful. Gruen et al. (2013) devised a programme for laboratory cats so that they were able voluntarily to leave their enclosure to enter a cat carrier, then be calmly transported to the laboratory, then cooperate with physical examination. The reward-based programme entailed the cat gradually becoming familiar with the laboratory technicians, then gentle handling and finally entering a carrier and being transported to the research facility.

While it is impossible to be sure of the exact number, the figures from insurance companies suggest that about 230,000 cats are killed annually in road traffic accidents; these accidents can also lead to life-changing consequences for the cat such as the loss of a leg. Such accidents are of obvious concern to owners both emotionally and financially in terms of vets' bills. Kasbaoui, Cooper, Mills, and Burman (2013) investigated the effectiveness of electronic containment systems for preventing traffic accidents involving cats as well as lowering the risk of other perils such as poisoning and disease. The study compared the welfare of 23 domestic cats contained by an electronic system for more than 12 months with 23 freely roaming cats with no experience of containment. Kasbaoui et al. reported that the cats' long-term quality of life is not harmed by the restrictions imposed by an electronic boundary fence.

From the animal's perspective, if not the human's, there is a clear contrast between the use of reward and punishment. What are the relative merits of the two approaches to training?

Reward or punish?

Rooney and Cowan (2011) asked 53 male and female dog owners about the type of training they had used with their dog and looked at the dog's current behaviour. All the owners said they employed both approaches although some favoured punishment, including physical punishment, while others were more likely to use reward. Rooney and Cowan found that when the owner favoured the use of punishment the dog was less likely to interact with a stranger, while dogs who were physically punished tended to be less playful. The dogs trained predominately with rewards and where the owner took a patient approach to training tended to perform better in a novel training task. Rooney and Cowan conclude that: "For dog owners, the use of reward-based training appears to be the most beneficial for the dog's welfare, since it is linked to enhanced learning and a balanced healthy dog–owner relationship" (p. 176). In a similar vein, Cooper, Cracknell, Hardiman, Wright, and Mills (2014) compared the effects on dogs of reward-based training and remote electronic training collars. In terms of the effectiveness of training, they reported no advantage to the use of collars but found that "The immediate effects of training with an e-collar give rise to behavioural signs of distress in pet dogs, particularly when used at high settings" (p. 11).

Are aversion-based training methods a risk to animal welfare? Fernandes, Olsson, and Vieira de Castro (2017) reviewed 14 studies and concluded that: "The existing research papers on the topic suggest a correlation between the use of aversive based training methods and indicators of compromised welfare and behavioural problems in dogs" (p. 10). However, they also point to several shortcomings in the literature: 1) a predominance of surveys rather than objective measures, 2) an over-representation of samples of police and laboratory dogs, and 3) a concentration on shock-collar training which is only one of several devices used in aversive-based training.

Greenebaum (2010) contrasts harsh obedience training that emphasises human dominance with reward-based methods. The former is *human-centric* aiming to place the dog in a subordinate position; the latter is *dog-centric* and encourages companionship rather than dominance-seeking to balance the needs of humans and dogs. Ziv (2017) reviewed 17 studies of the effects of different approaches to dog training, including positive reinforcement, positive punishment, and other aversive procedures, on the dog's behaviour towards humans and other dogs, its physiological functioning, and its welfare. Ziv concluded that aversive training methods, such as positive punishment and negative reinforcement, pose a risk to the dog's physical and behavioural well-being. The point is also made that while positive punishment can be an effective method, there is no indication that it is any more effective than training based on positive reinforcement.

Makowska (2018) prepared a comprehensive review of dog training methods for the British Columbia Society for the Prevention of Cruelty to Animals (BC SPCA) to inform the development of humane standards for dog trainers in British Columbia. Makowska notes a strong advocacy for reward-based training across many international animal welfare agencies, including The RSPCA and The Kennel Club, alongside an opposition to aversive methods. Makowska makes reference to training techniques called "hanging and helicoptering", hanging takes the form of lifting the dog from the ground by its collar and helicoptering the dog when it is lifted and spun round by its collar. Grohmann, Dickomeit, Schmidt, and Kramer (2013) record the case of a 1-year old German Shepard whose owner had used hanging as a training method. The dog had profound brain injuries consistent with strangulation and had to be euthanised. There is a fine line between punitive training and animal abuse and a strong argument could be made that hanging and helicoptering crosses that line.

The paradox within aversive training methods is twofold: 1) removing an unwanted behaviour does not guarantee that a desired behaviour will appear – how could the animal possibly know it is required to do in place of the problem behaviour? and 2) causing suffering may well have an emotional impact on the animal, so that it becomes unduly fearful or aggressive, which is decidedly not the way to bring about a positive owner–animal relationship. Seligman's research, discussed in Chapter 1, demonstrates the detrimental effects of unavoidable punishment on dogs. As may be anticipated, the type of training can enhance or diminish the level of attachment between dog and owner (Vieira de Castro, Barrett, de Sousac, & Olsson, 2019).

If the use of aversive training methods is contra-indicated why, with commercial support, do they continue to be used?

Why favour punishment?

Todd (2018) considers the barriers faced in convincing the general public to use humane dog training methods. Todd lists a lack of understanding of the welfare risks inherent in punitive methods, the poor quality of a great deal of relevant information available to those dog owners who look for it, a lack of regulation of dog trainers, and low levels of theoretical and practical knowledge about dog training. The use of punishment to train animals may be ingrained into people's understanding of social norms so that, as Todd states, "Because people frequently cite themselves as the source of their dog training knowledge, it may be difficult to reach them with messages about appropriate training methods" (p. 32).

To expand on Todd's position, punishment in training animals may be an instance of wider ranging beliefs about the appropriateness of punishment. A familiar defence of punishment is that a failure to punish – as evident in the saying "spare the rod and spoil the child" – is a mistake. While punitive views are sometimes associated with religious belief (Beller, Kröger, & Kliem, in press; Laurin, Shariff, Henrich, & Kay, 2012) the aphorism "spare the rod and spoil the child" is neither of Christian nor Biblical origin. The phrase actually comes from a narrative poem titled *Hudibras* written in 1684 by Samuel Butler. Nonetheless, there is a widespread belief that punishment is justifiable as seen, for example, in hitting children who misbehave. Although there is robust evidence that physical punishment is detrimental to child development (Gershoff et al., 2018), the practice of hitting children persists.

It is highly unlikely that the use of punishment to change behaviour will disappear, but steps can be taken to reduce its incidence. In an increasing number of countries it is illegal to hit a child; while several professional bodies express condemnation of punitive training methods for animals. The European Society of Veterinary Clinical Ethology (ESVCA; Masson et al., 2018) takes the position that: "Members of ESVCE strongly oppose the use of e-collars in dog training ... and we urge all European countries to take an interest in and position on this welfare matter" (p. 75). The views of the ESVCE are echoed in the guidelines on canine and feline behaviour management published by the American Animal Hospital Association (AAHA; Hammerle et al., 2015).

The arguments for the use of training based on reward, not punishment, are not exclusive to companion animals. This point is supported by studies of the effectiveness of training regimes for both ponies (Torcivia and McDonnell, 2018) and horses (Baragli, Mariti, Petri, De Giorgio, & Sighieri, 2011).

Reducing aggression

There are many instances in everyday life where it is desirable to reduce the likelihood that some harmful occurrence will take place. Two broad strategies can be used to achieve this aim: the first, *prohibition*, seeks to prevent harm by eliminating, or at least largely reducing, the possibility of exposure to the cause of the harm. This prohibitive approach may entail legislation and subsequent enforcement as seen with, for example, the requirement to use seatbelts in cars to reduce serious injury or banning smoking in public places to safeguard public health. The second approach, *contingency management*, aims to reduce harm by changing the behaviour of those involved in the harmful event.

Prohibition

When it comes to reducing the harm caused by companion animals, a radical approach would be to prohibit keeping a pet that has the potential to cause serious injury. In some countries, there are laws stating which animals may be privately owned and specifying the conditions under which they must be kept. In the UK, it is illegal to keep a range of animals, from aardvarks to zebras, without a licence (see Department for Environment, Food and Rural Affairs, 2007).

In some countries in the developing world, it is necessary to control dogs because they carry diseases such as rabies and toxocariasis and so present a risk to public health (Reese, 2005).

The presence of certain animals within a household may be frowned upon for religious reasons. Menache (1997) suggests that in some religions, canines are perceived as posing "A threat to the authority of the clergy and indeed, of God" (p. 23). However, there are shades of grey: Fuseini, Knowles, Hadley, and Wotton (2017) note that:

> There is disagreement among Islamic jurists regarding the keeping of dogs as pets. Some Islamic jurists are of the view that dogs are ritually "dirty" animals and cannot even be kept as pets, they, however, approve the use of dogs in hunting and guarding. The Prophet is also reported to have said that Angels do not enter houses where dogs are kept. Others are of the view that classifying dogs as "dirty" is not Islamic, but a pre-Islamic Arabian culture. (p. 2)

It is much more common to take a targeted, rather than all-encompassing, approach in the use of legislation to attempt to prohibit aggressive dogs. A prime example of this line of action is seen where laws are enacted to target specific breeds of dogs.

The Dangerous Dogs Act 1991

While serious injury from a dog attack is comparatively rare, with death even more uncommon, when such cases do occur, they can evoke a strong public reaction. A public outcry, reinforced by the mass media, may prompt politicians to take legislative action. After a series of attacks by uncontrolled dogs, which included four deaths, *The Dangerous Dogs Act 1991* was introduced in England and Wales. There is similar legislation in Scotland and Northern Ireland. The Dangerous Dogs Act made it illegal to own a *specially controlled dog* without a court exemption (see Bennett, 2016). Those owners permitted to keep the four breeds of dangerous dogs specified by the Act – Pit Bull Terrier, Japanese Tosa, Dogo Argentino, and Fila Brasileiro – must take responsibility for ensuring that the dog is registered, insured, and always muzzled and kept on a leash when in public. In addition, the specified breeds must be neutered or spayed to prevent breeding as well as being tattooed and microchipped so that should the dog escape or be abandoned, its owner can be identified. The initial evidence suggests that microchipping has proved a successful strategy (Siettou, 2019). Finally, the Act prohibits all breeding, sale, and exchange of these dogs. Those who disobey the law face a penalty of 6 months in prison and/or a £1,000 fine.

In 1997, the legislation was amended (see Department for Environment, Food and Rural Affairs, 2009) with the introduction of *Dog Control Orders,* which replaced local byelaws and empowered Local Authorities to instruct owners to control their dogs in designated spaces and to pick up their dog's waste. Those owners who breach a *Dog Control Order* face a maximum fine of up to £1,000. The 1997 amendment also removed the requirement for mandatory destruction of dangerous dogs – between 1991 and 1997 an estimated 1,000 dogs were destroyed – allowing the judge's discretion on how best to deal with a dangerous dog.

The political motivation behind the creation of identifiable "dangerous dogs" can be understood in several ways. The most obvious is public protection, but in truth the risk posed to the public by these dogs is small in comparison to other risks, such as air pollution and excessive alcohol consumption, which do not demand the same legislative urgency. McCarthy (2016) points out that political concern about dangerous dogs stretches back to the early 1800s and panics about stray dogs, attacks on people and the presence of rabies. As McCarthy explains, these "mad dog" panics were linked to the impoverished living conditions of the working class and their favoured breeds of dog: a point which illustrates how a dog's breed can be symbolic of the owner's social class and status as illustrated by McCarthy's quotation from a national newspaper:

> Why is it that every time one sees a Staffordshire Bull Terrier walking down the street, the chances are the man at the other end of the lead has an IQ at sub-moronic level and a swagger that suggests an undeserved level of personal confidence? I've nothing against the dogs, but many of their owners are a waste of oxygen. (p. 568)

McCarthy notes how the language used in the newspaper article – the swagger, the undeserved personal confidence, an IQ at sub-moronic level – paints a picture of certain dog owners and their dogs through a tainted perspective of social class.

The theme of a moral panic is also used by Hallsworth (2011) in relation to the Dangerous Dogs Act 1991 and one specific breed of dog – the Pit Bull Terrier – which became the demon at the centre of the storm. Hallsworth is clear as to the prevailing situation: "Since the late 1980s, when the Pit Bull first came to public attention, this canine has found itself the object of an official campaign that has as its stated aim the wholesale destruction of the Pit-Bull as a breed. Not to put too fine a point on it, Britain's very own attempt at a canine genocide" (p. 391). How is it possible that one type of dog can invoke such feelings in a country that professes to be a nation of dog lovers (Figure 5.2)?

Hallsworth puts the case that the public image of the Pit Bull is the end product of a period of frenzied misinformation and panic fuelled not by hard evidence but by the politicians of the day in combination with the mass media. Thus, the conditions were created for a national moral panic where the atypical appears to be the norm, the exception is the rule and a small matter is grossly magnified in a glare of publicity. The Pit Bull became to be condemned as vermin, and American vermin at that, to be set apart from our own treasured pet dogs. The owners of these Pit Bulls used them as a symbol of their own tough image – hence the pejorative term *status dog* used for those breeds included in the Dangerous Dogs Act. Maher and Pierpoint (2011) and Hughes, Maher, and Lawson (2011) found that young people did own these dogs to enhance their status but that was not the whole story. In the same way as many other dog owners, the status dogs were kept for companionship, protection, and creating opportunities to socialise.

At one point, the Pit Bull became labelled, again without sustainable evidence and even supported by the Royal Society for the Prevention of Cruelty to Animals (2010), as a "weapon dog" trained to kill people and animals (see Hallsworth for a discussion of the RSPCA's non-defence of its position). As is generally the case with moral panics, the "evidence trail" is circular: the press know that Pit Bulls are dangerous because the

Figure 5.2 Pit Bull Terrier.
Source: Photograph by Angela Cavina. Pexels Licence: Free for personal and commercial use.

politicians say so, the police know because the politicians say so, the RSPCA know because … and so on *ad infinitum*.

A problem with breed specific legislation (BSL) is twofold as the American Veterinary Society of Animal Behavior (2014) explains: 1) "Breed alone is not predictive of the risk of aggressive behavior. Dogs and owners must be evaluated individually" (p. 1); 2) "An additional concern regarding BSL involves accurately identifying breeds or mixes that presumably fall under the restrictions. Visual identification is not reliable." (p. 2). The second point, supported by a study showing inconsistent identification of Pit Bulls by staff in dog shelters (Olson et al., 2015), casts doubts over the accuracy of newspaper reports about dangerous dogs which rely solely on witness identification.

Legislation against dangerous dogs was not confined to the UK, although we were in the vanguard. Australia implemented the British Act (without any instances of attacks by a Pit Bull, as acknowledged in a report published by the Parliament of the State of Victoria in 2016) while similar legislation was enacted in several other European countries including Denmark, France, Germany (Schalke, Ott, von Gaertner, Hackbarth, & Mittmann, 2008), Italy (Mariti, Ciceroni, & Sighieri, 2015), and the Nordic countries (Lie, 2017; Sarenbo, 2019). The Dutch government later repealed its breed specific legislation as did Lower Saxony, Germany (Ott, Schalke, von Gaertner, & Hackbarth, 2008), and Demark (Nilson, Damsager, Lauritsen, & Bonander, 2018).

Another characteristic of a moral panic is that it fades quickly after its day in the sun (Goode & Ben-Yehuda, 1994) sometimes to be replaced by a new demon. Nonetheless, when a dog is given a bad name, it can be difficult to lose the corresponding public image: an Italian study by Gazzano, Zilocchi, Massoni, and Mariti (2013) found that Pit Bull terriers continued to evoke more fear than other types and sizes of dogs.

Local control

The use of legislation on a national scale may be a blunt way to achieve lower rates of dog bites with better success possible with targeted interventions on a local scale. A Canadian study reported by Clarke and Fraser (2013) reviewed animal control procedures – including rates of dog licensing and ticketing (violation notices issued by animal control enforcement staff) and local budgets – in 36 urban municipalities set against the rate of reported dog bites. The reported frequency of dog bites was generally higher in those municipalities which had higher licensing and ticketing as well as more staffing and greater budget. Where ticketing was at a very high level, the reported bite rate was much lower than predicted indicating a positive effect of enforcement. However, whether or not a municipality had breed-specific legislation did not influence the reported bite rate.

Vertalka, Reese, Wilkins, and Pizarro (2018) reported that in the city of Detroit, the traditional predictors of dog bites – crime, building vacancy, and building blight (the presence of rodents and trash) – functioned differently in different parts of the city. Those neighbourhoods with high levels of property vacancy and building blight had higher rates of bites, probably because they are territories for stray and feral dogs, as did the areas of the city with amenities that attract children, such as schools and parks. Those areas high in commercial and retail venues had a reduced rate. Vertalka et al. suggest that while preventative programmes (see below) may be usefully targeted at areas where children are concentrated but other measures are required elsewhere in the city.

Contingency management

If Skinner's notion of a three-term contingency (see Chapter 1) is applied to dog attacks, the first step is to consider the antecedents to the aggressive behaviour. Some antecedents, such as the breed of dog, are *static* in that they cannot be changed; however, other antecedents, such as the interaction between the child and the dog, are *dynamic* and so allow the possibility of change. In a given incident, there are three interacting sets of behaviour prior to the bite to consider: these are the actions of the owner, the victim, and the dog.

The owner

It can be difficult to pinpoint exactly what constitutes aggressive behaviour. While there is little doubt at the more extreme end of the scale it may not be so straighthood to decide when a puppy's playfulness blends into aggressive behaviour. Yet further, not every owner will have the same criteria for what constitutes aggression nor react to control it in the same way. An Italian study showed that not all owners will attend to their dog's behaviour in public even when they are aware that it constitutes a problem (Mongillo, Adamelli, Pitteri, & Marinelli, 2015). Orritt, Gross, and Hogue (2015) showed that the attitudes of dog owners to aggression were associated with their levels of experience with dogs, including being able to read the signs of aggressive behaviour. Orritt, Gross, and Hogue found that some owners felt it is always easy to blame the dog and that sometimes it is necessary to take the dog's side.

As discussed previously, some people have a particular breed of dog because of the status they perceive comes with ownership (Maher & Pierpoint, 2011). What are the psychological characteristics of those people who choose to have an aggressive breed of dog? Wells and Hepper (2012) used the Eysenck Personality Questionnaire (EPQ; Eysenck & Eysenck, 1964) to compare the personalities of male and female owners of aggressive (German Shepherd, Rottweiler) and non-aggressive (Labrador Retriever, Golden Retriever) dogs. The EPQ measures the three personality dimensions of Extraversion (E), Neuroticism (N), and Psychoticism (P): E corresponds to an individual's level of sociability; N denotes degree of emotional stability; and P was originally seen as the personality factor underlying psychosis but was later described as more akin to psychopathy (Eysenck & Eysenck, 1972). Individuals with higher P scores are more likely to be irresponsible, to break social norms, and to be aggressive. Wells and Hepper found that while E and N had minimal effect, the owners of the aggressive breeds had significantly higher P scores than the owners of the non-aggressive breeds.

As Grigg (2019) points out, the owners of dogs displaying problematic behaviour may have some difficulties in accepting the need for change which, in turn, is related to their personality and to their dog's characteristics (Turcsán, Range, Virányi, Miklósi, & Kubinyi, 2012). The use of formal personality testing may help professionals working with owners and their aggressive dog to adjust the focus of their work guided by the potential importance of personality factors.

The victim

In looking at the context in which dog bites occur Oxley, Christley, and Westgarth (2018) make the observation that "A wide variety of situations were noted with the most common being the victim involved with some form of direct interaction with the dog. In particular, stroking or attempting to stroke the dog was the most commonly reported. Playing with the dog or handling/lifting/restraining the dog were also key bite contexts" (p. 35). In a similar vein, Arhant, Beetz, and Troxler (2017) found that a child's interactions with a dog often preceded a bite. These interactions were both benign, as in petting and stroking, and aversive to the dog as with hitting and removing its food. Rezac, Rezac, and Slama (2015) found that bites to the face were preceded by the person bending over the dog, putting their face close to the dog's face, and prolonged gazing between person and dog.

The dog

Like many species, including humans, a dog communicates through a range of non-verbal signals. These signals, sometime singly sometimes in combination, have discrete functions such as showing aggression, indicating a willingness to begin an interaction, appeasement to another animal, and to calm an aggressive encounter. These signals involve, among others, variations of posture, degree and type of head and body movement, length of eye contact, licking of lips or nose, sniffing, and tail position. Thus, for example, a dog may seek to calm an interaction with another dog by turning its head, licking its nose, freezing, and turning its body away (Mariti et al., 2017); or lick its lips and look away when signalling appeasement to humans (Firnkes, Bartels, Bidoli, & Erhard, 2017). Of course, the use of signals to communicate is not restricted to canines: Tibbetts (2013) discusses the use of a range of signals, from plumage displays to type of call or song, used by different animals for purposes such as conveying fighting ability to deter potential opponents and so minimise the costs of open conflict. To learn to attend and respond appropriately to the flow of social information in their environment is a basic developmental task for human children (Dodge, 1986).

Intra-species communication relies on a shared understanding of non-verbal signals so when it comes to inter-species communication, can humans comprehend what their pet is telling them non-verbally? Wan, Bolger, and Champagne (2012) showed videotaped recordings of dogs, previously judged by experts in dog behaviour as showing either a happy or a fearful dog, to sample of over 2,000 adults with various levels of experience with dogs. The observer's level of dog experience was a significant predictor of correct identification some of the dog's emotions: the probability of correct identification of a fearful dog was 0.30 for adults with no history of living with a dog but greater than 0.70 among experienced adults. However, for a happy dog, both experienced (0.93) and inexperienced (0.90) adults make accurate identifications. A difference in cue utilisation was evident, with experienced adults observing more of the dog's physical features, particularly the ears, to inform their judgement.

Lakestani, Donaldson, and Waran (2014) reported a similar study in which children (aged 4 to 10 years) and adults watched videos of dogs behaving in a friendly, aggressive, or fearful manner. All the age groups could identify friendly and aggressive behaviour but children under 6 years of age were poor at identifying fear. The misidentifications of the dog's fearful behaviour were accounted for by the observer watching just one feature, typically children looking at the dog's face, rather than attending to a combination of features. Aldridge and Rose (2019) reported that while young children said they would avoid an angry dog, they did not express any qualms about approaching a frightened dog. Yet further, from the view of bite prevention, adults can also have difficulties in accurately interpreting a dog's non-verbal behaviour when it is interacting with a child (Demirbas et al., 2016).

As an aggressive encounter may be preceded by specific signals by which the dog seeks to communicate its intentions, it may be possible to educate people to understand dog behaviour to reduce the likelihood of an aggressive incident (Orritt et al., 2015). The low baseline of knowledge about dog bite prevention in children and their parents and guardians (Dixon, Mahabee-Gittens, Hart, & Lindsell, 2012) and the many costs of injuries caused by dog bites, particularly to children, led to the development of prevention programmes.

Prevention programmes

Prevention programmes often take an age-appropriate educational approach to teach children to recognise dog emotions and to behave safely when with dogs. These programmes are delivered in a manner similar to a school lesson with workbooks, videotaped examples to discuss and homework tasks. In a survey of 34 paediatric victims and their families, after a dog bite there was a majority view that children may benefit from interventions to assist with post-bite fears and that families would gain from learning more about bite prevention (Boat, Dixon, Pearl, Thieken, & Bucher, 2012). There have been steps in this direction with the development of several bite prevention programmes including *BARK* (Be Aware, Responsible, and Kind), *Prevent-a-Bite*, and *Delta DogSafe*: Lakestani and Donaldson (2015) give an overview of these programmes alongside an evaluation of their own educational intervention with preschool children.

An Australian study by Chapman, Cornwall, Righetti, and Sung (2000) used a randomised design to evaluate the "PreventaBite" programme for primary schoolchildren. They found that the programme was successful in the short term as the children increased their defensive behaviour around strange dogs. However, Chapman et al. caution that short-term gain does not necessarily become long-term change: Will the effects endure or will "booster" sessions be required? Further, will the gains reduce the number of children bitten by dogs? A study conducted in China by Shen, Pang, and Schwebel (2016) also utilised a randomised design in the evaluation of the effects on children of watching an educational video of testimonials (verbal descriptions of the experience of having been bitten) on dog-bite prevention. The intervention was successful in that compared to a control group, the children who watched the testimonials showed a significantly higher level of knowledge about safety and vulnerability in interacting with dogs. In a simulated exercise with dogs, the children demonstrated significantly less risky behaviour.

The Blue Dog Programme

As described by Meints and de Keuster (2009), the *Blue Dog Programme* was designed for use with male and female children under the age of 7 years. The programme, presented on a compact disc and printed booklet, allows the child to view scenes of interactions, with safe and unsafe outcomes, between cartoon dogs (the blue dogs) and cartoon children. The children then make decisions about a cartoon child's interactions with the dog. In addition, the homework involves parents in assisting their children to evaluate the scenarios and choose their responses. As well as helping the children learn, it is anticipated that parental engagement will highlight and inform parents of their child's level of knowledge and supervision requirements. In addition, parents are given an instruction booklet with details of effective education and supervision regarding their child's safety with dogs. Meints and de Keuster reported a preliminary evaluation showing that the programme was effective in teaching children about safe behaviour with dogs.

Schwebel, Morrongiello, Davis, Stewart, and Bell (2012) reported an evaluation of the Blue Dog Programme carried out with 76 children aged from 3.5 to 6 years. The evaluation used a robust methodology, a pre-post randomised design. Schwebel et al. summarise their findings: "The present results indicate that some knowledge may be transferred to children, but that knowledge is not necessarily utilized effectively by children when they encounter dogs" (p. 278). Morrongiello et al. (2013) employed a similar randomised design to Schwebel et al. in focussing on parental reactions to their child's behaviour when they were in close proximity to a strange dog. The findings were not as expected, Morrongiello et al. concluded: "In fact, surprisingly, the results suggest several ways that parents may actually contribute to elevate children's risk of experiencing a dog bite by unfamiliar dogs" (p. 112). Arhant, Landenberger, Beetz, and Troxler (2016) similarly found that some parents had a poor understanding of animal behaviour in the context of parent-child interactions.

As in other evaluations of attempts to change social behaviour, such as social skills training (Hollin & Trower, 1986), the evaluation of dog bite prevention programmes hinges on two fundamental questions: 1) does the programme effectively impart the necessary knowledge and skills and 2) do the newly acquired knowledge and skills produce the intended outcome? The answer to these two questions relies on disentangling *process* and *outcome*. The extant evidence gathered using robust research designs allows confidence in stating that the process of conducting a programme is effective in that children do gain in knowledge. However, it is less clear that the programme gains generalise to real life and actually reduce the number of children who are bitten. While process is important and continues to be refined (Meints, Brelsford, & De Keuster, 2018), the reviews emphasise the need for robust outcome evidence to inform programme improvement (Duperrex, Blackhall, Burri, & Jeannot, 2009; Shen et al., 2017).

While bite prevention programmes have a role to play, they are not a panacea. Westgarth and Watkins (2015) conducted in-depth interviews with eight women who had been bitten by a dog. The interviews revealed that in some instances there was no interaction with the dog prior to the bite so it would have been impossible to read the dog's behaviour and take appropriate action. In such cases dog bites are clearly not preventable through educational programmes with a focus on canine non-verbal communication. Westgarth and Watkins observed that some women said they believed, "it would not happen to me" which connects with the wider psychological context of people's views on accidents. As Hemenway (2013) points out, there are psychological barriers to people seeing the relevance of accident prevention. There are widely held beliefs that the event will never happen to them, or that accidents happen and that's just the way it is, or that in a just world the accident was probably deserved.

Reducing anxiety

The strategies used to reduce anxiety are broadly comparable to those used to reduce aggression, although in practice more directly focused on the animal's rather than the owner's behaviour. The biological interventions found in the treatment of anxiety are focused on a range of drugs.

Drug regimes

There are various classes of drugs, with new varieties being refined, used to treat anxiety in dogs: Beata et al. (2007a) describe *anxiolytics*, such as beta-blockers and benzodiazepines which act as a sedative; *antidepressants*, including tricyclic antidepressants and selective serotonin reuptake inhibitors (SSRIs); *dog appeasing pheromone;* and *neutriceuticals*, such as alpha-casozepine and L-Theanine. Some anxiety-reducing drugs are used generically across a range of disorders (e.g., Pineda, Anzola, Olivares, & Ibáñez, 2014) and others with a specific type of anxiety disorder (e.g., Cannas et al., 2014). Some drugs may be used across species: a French study by Beata et al. (2007a) used alpha-casozepine to treat anxious dogs following the same successful use of the same drug with anxious cats (Beata et al., 2007b). In the previous chapter, the case of Milo, a *Chiweenie* with a fear of thunderstorms, was discussed (Radosta, 2016). A treatment regime combined drugs (an anxiolytic) and relaxation training enabling Milo to cope with storms.

Drugs may also be used in combination: Pineda et al. (2014) used clorazepate dipotassium (a serotonin reuptake inhibitor) together with fluoxetine (a benzodiazepine) with 36 dogs with a range of anxiety disorders. They make the point that the drugs should be combined with a behaviour modification programme; in this case, the programme involved the owners teaching the dog to be relaxed when it could become anxious and reinforcing the dog when it is relaxed.

In the study cited above, Pineda et al. reported that the drug plus behaviour modification programme was effective for 25 of the 36 dogs. The dogs which did not benefit were primarily those that were aggressive, indeed the behaviour of these dogs worsened; this untoward effect indicates, as Pineda et al. suggest, that in such cases this approach may constitute a risk.

Obsessive compulsive behaviour manifest as spraying, overgrooming, and self-harm can be a problem for both dogs and cats (Overall & Dunham, 2002). An Australian study by Seksel and Lindeman (1998) used clomipramine (an antidepressant) to treat 11 cats with obsessive- compulsive disorder: seven cats sprayed urine, there was overgrooming in three cases, and excessive vocalisation in one cat. The drug regime was accompanied by a behaviour modification programme in which the cat was discouraged from spraying by for example making the litter tray more attractive and encouraging the cat to groom when it was not anxious. The combination of drug and behaviour modification was successful in 10 cases. However, Seksel and Lindeman note the problem of disentangling the effect of the drug from the behaviour management: "Ideally, a double-blind study comparing clomipramine alone, behaviour modification alone and both therapies together would be undertaken to assess operator effect and identify any placebo effect" (p. 320).

In a later study, Seksel and Lindeman (2001) used clomipramine to treat 24 dogs variously displaying obsessive-compulsive disorder, separation anxiety, and noise phobia. Alongside the drug, which was gradually withdrawn over time, a behavioural programme was followed that included avoiding leaving the dog alone, rewarding relaxed behaviour, rewarding short periods spent alone, and no punishment. Seksel and Lindeman reported that 20 of the dogs showed a significant to moderate improvement, some to the extent where the problematic behaviour disappeared. Seksel and Lindeman also noted the poor implementation of the behavioural programme by some owners (see below).

Cannas, Frank, Minero, Godbout, and Palestrini (2010) used clomipramine with 23 dogs who became distressed, barking excessively and acting destructively, when left alone at home. The drug treatment was accompanied by a behaviour modification programme in which the owners were taught to read their dog's non-verbal communication, particularly with respect to anxiety, and to reinforce the dog's desirable behaviour while undesirable behaviours were ignored or interrupted. Cannas et al. found that treatment reduced the signs of anxiety and no harmful reactions to the drug were evident.

Michelazzi et al. (2015) used the drug L-Theanine in treating 20 dogs with a noise phobia. L-Theanine is an amino acid present in tea leaves and is commercially available, for people as well as animals, as an aid to relaxation with the benefit of not causing drowsiness. The dogs were divided into two groups: the drug was administered to ten dogs but not to the ten dogs in the control group. All 20 dog owners were given identical instructions on managing the dog's behaviour such as ignoring phobic behaviours and demands for attention while rewarding relaxed behaviours. This study has a strong design as it allows inferences to be drawn about the role of the behavioural component. The results showed an improvement in the phobic behaviours of *both* the treatment and control groups. Michelazzi et al. reach the view that behaviour management "Is to be considered essential for the treatment of noise phobias" (p. 57).

Landsberg et al. (2015) explain the rationale underlying the use of *pheromones* to treat anxiety: "The dog-appeasing pheromone (DAP) ... is a synthetic analogue of the pheromone secreted after parturition by the intermammary sebaceous glands of the lactating bitch (Pageat & Gaultier, 2003). This pheromone is responsible for the sense of wellbeing experienced by puppies when with their mother" (p. 260).

Sheppard and Mills (2003) used pheromones in the treatment of 30 dogs that displayed a fear of fireworks. The pheromone was administered via an electric plug-in diffuser placed close to the dog's usual resting place over the "Guy Fawkes period" when fireworks are at their zenith. The owners were also given guidelines for managing fearful behaviour such as not punishing the dog and keeping it in a secure safe place. Sheppard and Mills reported that some fearful behaviours, including barking, cowering, and trembling, were successfully ameliorated by the treatment. However, some other behaviours, such as soiling and being startled, did not change, which may be a baseline effect due to their low overall frequency. Levine and Mills (2008) found that the combination of pheromone and behavioural treatment for dogs with a fear of fireworks proved effective at a 12-month follow-up.

The use of pheromone collars is not restricted to dogs. DePorter, Bledsoe, Beck, and Ollivier (2019) used a plug-in pheromone diffuser in a programme aimed to reduce aggression in 45 multi-cat households. Before the diffuser treatment started, the owners received directions for effective management of aggressive incidents which emphasised positive reinforcement and strongly discouraged punishment. The pheromone treatment was used in 20 households and a placebo in 25 households and the frequency and intensity of aggressive interactions were monitored. The behaviour management directions appeared to have an immediate effect in reducing aggression even before the introduction of the pheromone diffuser. DePorter et al. conclude that "Pheromones may be useful as a component of a complete behavior modification program" (p. 304).

The administration of the pheromone by a diffuser requires the treatment takes place indoors. When the dog is outdoors a *pheromone collar* provides an alternative. A plastic collar embedded with pheromones is placed around the dog's neck allowing the dog's natural body heat to release the odourless pheromone. Landsberg et al. tested the effects of a pheromone collar on anxiety caused by sound using 24 beagles divided into treatment and placebo groups. They found that compared to the placebo group, the dogs with pheromone collar improved on ratings of fear and anxiety.

The advantages of collars and diffusers is that they are not intrusive or painful as would be the case with an injection. The drugs can also be introduced in the animal's diet. Landsberg, Milgram, Mougeot, Kelly, and de Rivera (2017) conducted an evaluation of a proprietary cat food containing the anxiety reducing drugs alpha-casozepine and L-tryptophan with 24 domestic cats of which 12 received the treatment and 12 formed a control group. The found that the diet was effective in reducing anxiety but this effect was limited to situations associated with moderate levels of anxiety. As the two drugs were administered simultaneously, it is not possible to say whether the effect they produce was due to one particular drug or to their combined effect.

Homeopathy

The notion of homeopathy dates back to the idea put forward by Samuel Hahnemann (1755–1843) that "like cures like." The basis of homeopathy is that whatever is responsible for the symptoms of a disease in healthy people will be a remedy for similar symptoms in ill people. This idea has survived through the centuries and is now one of a number of alternative or complementary medicines such as acupuncture, dietary supplements, probiotics and tai chi. A British study by Cracknell and Mills (2008) provides an example of homeopathic treatment with 75 dogs with a strong fear of fireworks.

Cracknell and Mills employed a strong research design, a double-blinded, placebo-controlled trial, as advocated by Seksel and Lindeman (1998), to evaluate the efficacy of a homeopathic remedy for the fear. The study was carried out over New Year when fireworks are in abundance. The dogs were randomly assigned to either a homeopathic treatment group receiving a "Potentised homeopathic remedy (verum), based on phosphorus, rhododendron, borax, theridion, and chamomilla (6C and 30C in 20% alcohol)" (p. 82) or to a placebo group given water and 20% alcohol in a bottle matching the treatment group. The owners were also encouraged to follow behavioural advice such as not punishing the dog when it is frightened, not reassuring the dog when it is frightened (risks reinforcing the fearful behaviour), and providing a safe and comfortable place as a refuge.

The evaluation was based on owners' ratings of their dog's fear, including changes in the 15 behaviours which they nominated as indicative of their dog's fear. In the homeopathic treatment group there was a significant improvement over the span of the study on all 15 of the owner's ratings of the signs of fear. In the placebo group, the owners' ratings indicated an improvement in 14 of the 15 signs of fear. In both groups there was also a significant improvement in the owners' rating of the global severity of their dog's responses. Thus, there was no significant difference *between* groups although there was a marked improvement *within* each group. As both groups improved the likely explanation for the change lies in the owners' use of the behavioural advice. Cracknell and Mills looked at this in detail and concluded that: "The piece of behavioural advice best adhered to across both treatment groups was *'Don't punish you dog when he is scared'* and *'Make sure your dog is kept in a safe and secure environment at all times'*, and it may be that these simple measures largely account for the effect seen" (p. 87).

The Cracknell and Mills study is a small detail within a very large and controversial picture: there is a wide-ranging debate, for both animals and humans, about the premise underlying homeopathy, the use of homeopathic medicines, and the state of the evidence. The debate around the use of homeopathy for animals is encapsulated by the diversity of views among those who express scientific doubt about the approach (e.g., Lees et al., 2017a, 2017b), those who express reservations about the reliability of the evidence (e.g., Mathie & Clausen, 2015a), and, once evidential bias is controlled, are in favour of its use (e.g., Mathie & Clausen, 2015b).

DePorter et al. (2019) make the point that drug treatment is most effective when accompanied by some form of behaviour management. It is feasible that the drugs are not necessary for successful treatment but it is more likely that the drugs create a window within which behaviour change can take place.

Non-drug treatment

Another approach to anxiety management lies in the use of an *anxiety wrap*, a commercial version which has the trade name *Storm Defender*®. Pekkin et al. (2016) explain that the application of pressure to the body can lower heart rate, change physiological functioning and reduce anxiety in both humans and animals. A pressure vest, or anxiety wrap, can be used with both anxious cats and dogs. The wrap looks like a dog coat, typically fitted around the neck and upper torso, and can be used to calm the animal at times of stress such as thunderstorms. In a trial of the pressure vest with 28 dogs, male and female, Pekkin et al. found positive effects in terms of mediating the effects of fireworks noise and speeding up recovery after stressful event.

Cottam and Dodman (2009) recruited 23 dog owners to take part in a comparison of the Storm Defender® cape (13 dogs) with a placebo cape (10 dogs) in reducing their dog's fear during thunderstorms. The Storm Defender® has anti-static properties as thunderstorms can produce a build-up of static electricity on the animal's fur which can discharge and cause shock. Cottam and Dodman reported that both capes resulted in a significant decrease in anxiety set against baseline measures.

Thus, 70% of owners using the Storm Defender® cape and 67% of the placebo cape owners reported an improvement in their dog's behaviour. They suggest that "a placebo effect or "deep pressure touch" are possible explanations for the owner-reported therapeutic effect" (p. 84).

Cottam, Dodman, and Ha (2013) used a pre-post design with the owners of 18 dogs afraid of thunderstorms to investigate the use of a product named *Anxiety Wrap*. This product is claimed to reduce a dog's fear of thunderstorms "By 2 pressure-inducing methods: maintained pressure 'swaddling' and acupressure" (p. 154). At the baseline phase each owner rated the intensity and duration of nine of their dog's anxiety-related behaviours (e.g., panting, shaking, and inappropriate elimination) during two thunderstorms. These scores were collated to give a mean Baseline Anxiety Score. The owners completed to same rating procedure as at baseline for five thunderstorms when they used the Anxiety Wrap. during five subsequent thunderstorms. Again, following the same procedure as at baseline, these scores formed five distinct Treatment Anxiety Scores. After the fifth use of the Anxiety Wrap, with no reported side effects, the mean anxiety score was significantly lower (by 47%) than at baseline; 89% of owners said the wrap was an effective treatment; and 80% of the owners said they would continue to use the wrap after the trial. There is also evidence to show that a wrap can be effective in the treatment of dogs with separation anxiety disorder and generalised anxiety disorder (King, Buffington, Smith, & Grandin, 2014).

Some treatments move away completely from drugs and pressure wraps in using standard behavioural methods of change. *Systematic desensitisation* is a classic treatment method used with human anxiety disorders. Developed by the psychiatrist Joseph Wolpe (1915–1997), the effects of systematic desensitisation are gained by the individual learning to cope with a graduated exposure to the fearful object or situation (Wolpe, 1958, 1962). Butler, Sargisson, and Elliffe (2011) used systematic desensitisation in the treatment of eight dogs with separation anxiety as seen in their destructive behaviour, soiling or excessive barking when alone. Butler, Sargisson, and Elliffe describe the treatment protocol:

> Owners carried out the treatment themselves following instructions provided. Owners were instructed to place their dog in isolation with food treats 3–4 times per day, with a minimum of 1 h between isolation periods. Starting with a 5-min separation period, owners were instructed to increase gradually and variably the period of separation in increments of 5 min until a period of 30–90 min was reached without recurrence of separation-related behaviour. After reaching that point, isolation durations were increased more rapidly. If the dog displayed evidence of separation-related behaviour, owners were instructed to return to the longest period not previously associated with separation-related behaviour and to proceed more gradually. (p. 140)

Butler, Sargisson, and Elliffe report that at completion of treatment the systematic desensitisation had reduced the severity and frequency of the problematic behaviours for all eight dogs. This positive effect was maintained at a 3-month follow-up.

In a review of approaches to the treatment and management of separation anxiety disorder, Sargisson (2014) notes that systematic desensitisation, sometimes supplemented by drugs, is a treatment of choice for canine separation anxiety disorder.

The literature suggests that pet problems can be successfully managed and positive improvements in behaviour achieved. There are obvious gains to be had, for both animal and owner, in successful treatment so what is the essence of success? The weight of evidence and professional experience highlights four elements to successful treatment.

First, treatment by drugs alone is not a means to lasting behaviour change. The outcome from the evaluations that drugs in combination with behavioural treatment are most effective highlights an important point for pet owners. When faced with a pet's difficult behaviour it is easy to explain the issue in physiological terms, to use a so-called "medical model" whereby biological functioning is the explanation for the problem. Of course, there are instances when the medical model is absolutely correct: an infection may be best treated with antibiotics, an allergy with antihistamines and so on. Thus, the individual takes their medicine, their condition improves and ultimately they are healthy once again. However, while behaviour has a biological element, it is much more than just a biological condition: 1) it is learned, 2) it has a psychological or individual element, and 3) it has environmental and social elements. The complexities of behaviour mean that behavioural change is not analogous to medical change. As well as any biological component, it is necessary to attend to the psychological and social elements to bring about change. This is as true for treating and training animals as it is to changing human behaviour.

Second, just like people, not all animals are equal with regard to the ease of behaviour change. Serpell and Hsu (2005) conducted a survey of the owners of 1,563 dogs that included 11 common breeds. The owners used a standardised questionnaire, the C-BARQ, to assess their dog's "trainability." The three breeds judged to be the most trainable were the Labrador Retriever, the Golden Retriever, and the Shetland Sheepdog; the three least trainable were the Dachshund, the Siberian Huskey, and the Bassett Hound. It follows that for all dogs some allowance will need to be made for how quickly and to what level training can have an effect.

Third, there are several ways to bring about behaviour change, some based on reward and encouragement, others on aversion and punishment. The overwhelming message from the research, endorsed by several professional bodies, is that positive methods are to be strongly encouraged for several reasons. The bottom line is that it is more effective to use reward-based rather than punishment-based methods to achieve a good outcome. However, as Alnot-Perronin (2005) notes, there are also practical and ethical issues in using pain to train an animal. An animal in pain can be unpredictable and dangerous with the risk of escalating aggression and the animal turning on the trainer. The trainer's motives for using punishment may be questionable: Alnot-Perronin refers to owners who physically beat their dog so as to "win" a confrontation. A problem here lies in the potential to cross the line from physical chastisement to abuse as seen in with "hanging," "helicoptering," and burning the dog with cigarettes. (A not dissimilar situation can arise when a parent or guardian's behaviour changes from psychical chastisement to physical abuse when attempting to control a non-compliant child.) The trauma associated with pain can be long-lasting so that punitive training methods may damage the animal over an extended period.

Fourth, in the human–animal training dyad, it is essential to pay attention to the owners as well as to the animals. When not implemented correctly even the best training regime may be doomed to failure. In the behaviour change literature the correct implementation and consistent adherence to a particular method is known as *treatment fidelity* or *treatment integrity* (Hollin, 1995). There are ample recorded instances of a lack of treatment integrity leading to a less than optimum outcome. For example, Seksel and Lindeman (2001) made the comment that "inadequate implementation" of the behavioural treatment may account for the cases where the dog failed to respond. In a similar vein, Blackwell, Casey, and Bradshaw (2016) noted that their programme was designed to prevent separation anxiety in newly rehomed shelter dogs: "The efficacy of this program was limited by the apparently poor compliance of owners in following the advice" (p. 18).

There is a trade-off to be had between a high level of oversight of the programme and close adherence to the training plan to ensure the best possible outcome set against the costs in both time and money this level of administration entails. This dilemma particularly applies to innovative methods of delivering training such as giving owners instructions on a CD so they can learn training techniques at their own rate in their own time (e.g., Levine, Ramos, & Mills, 2007). A CD has the potential to reach a large number of owners at a reasonable cost but once started there is no knowledge about how the intervention is delivered and its eventual effectiveness.

References

Aldridge, G. L., & Rose, S. E. (2019). Young children's interpretation of dogs' emotions and their intentions to approach happy, angry, and frightened dogs. *Anthrozoös, 32,* 361–374.

Alnot-Perronin, M. (2005). Inappropriate use of pain as punishment in canine aggression toward household members. In Mills, D., Levine, E., Landsberg, G., Horwitz, D., Duxbury, M., Mertens, P. Willard, J. (Eds.), *Current issues and research in veterinary behavioral medicine: Papers presented at the 5th international veterinary behavior meeting* (pp. 232–235). West Lafayette, IN: Purdue University Press.

American Veterinary Society of Animal Behavior. (2014). *Position statement on breed-specific legislation.* Schaumburg, IL: AVMA.

Arhant, C., Beetz, A. M., & Troxler, J. (2017). Caregiver reports of interactions between children up to 6 years and their family dog — implications for dog bite prevention. *Frontiers in Veterinary Science, 4,* Article 130.

Arhant, C., Landenberger, R., Beetz, A., & Troxler, J. (2016). Attitudes of caregivers to supervision of child–family dog interactions in children up to 6 years—An exploratory study. *Journal of Veterinary Behavior, 14,* 10–16.

Bailey, J. (2018). Does the stress of laboratory life and experimentation on animals adversely affect research data? A critical review. *Alternatives to Laboratory Animals, 46,* 291–305.

Baragli, P., Mariti, C., Petri, L., De Giorgio, F., & Sighieri, C. (2011). Does attention make the difference? Horses' response to human stimulus after 2 different training strategies. *Journal of Veterinary Behavior, 6,* 31–38.

Beata, C., Beaumont-Graff, E., Coll, V., Cordel, J., Marion, M., Massal, N., & Marlois, N. (2007b). Effect of alpha-casozepine (Zylkene) on anxiety in cats. *Journal of Veterinary Behavior: Clinical Applications and Research, 2,* 40–46.

Beata, C., Beaumont-Graff, E., Diaz, C., Marion, M., Massal, N., Marlois, N.,… & Lefranc, C. (2007a). Effects of alpha-casozepine (Zylkene) versus selegiline hydrochloride (Selgian, Anipryl) on anxiety disorders in dogs. *Journal of Veterinary Behavior, 2,* 175–183.

Beller, J., Kröger, C., & Kliem, S. (in press). Slapping them into heaven? Individual and social religiosity, religious fundamentalism, and belief in heaven and hell as predictors of support for corporal punishment. *Journal of Interpersonal Violence.*

Bennett, O. (2016). *Dangerous dogs.* Briefing Paper Number 4348. London: House of Commons Library.

Blackwell, E. J., Bolster, C., Richards, G., Loftus, B. A., & Casey, R. A. (2012). The use of electronic collars for training domestic dogs: Estimated prevalence, reasons and risk factors for use, and owner perceived success as compared to other training methods. *BMC Veterinary Research, 8,* 93.

Blackwell, E. J., Casey, R. A., & Bradshaw, J. W. S. (2016). Efficacy of written behavioral advice for separation-related behavior problems in dogs newly adopted from a rehoming center. *Journal of Veterinary Behavior, 12,* 13–19.

Blackwell, E. J., Twells, C., Seawright, A., & Casey, R. A. (2008). The relationship between training methods and the occurrence of behavior problems, as reported by owners, in a population of domestic dogs. *Journal of Veterinary Behavior, 3,* 207–217.

Boat, B. W., Dixon, C. A., Pearl, E., Thieken, L., & Bucher, S. E. (2012). Pediatric dog bite victims: A need for a continuum of care. *Clinical Pediatrics, 51,* 473–477.

Brent, L., & Kutzler, M. (2018, June). Alternatives to traditional spay and neuter – Evolving best practices in dog sterilization. *Innovative Veterinary Care* (available at ivcjournal.com).

Butler, R., Sargisson, R. J., & Elliffe, D. (2011). The efficacy of systematic desensitization for treating the separation-related problem behaviour of domestic dogs. *Applied Animal Behaviour Science, 129,* 136–145.

Cannas, S., Frank, D., Minero, M., Aspesi, A., Benedetti, R., & Palestrini, C. (2014). Video analysis of dogs suffering from anxiety when left home alone and treated with clomipramine. *Journal of Veterinary Behavior, 9,* 50–57.

Cannas, S., Frank, D., Minero, M., Godbout, M., & Palestrini, C. (2010). Puppy behavior when left home alone: Changes during the first few months after adoption. *Journal of Veterinary Behavior, 5,* 94–100.

Chapman, S., Cornwall, J., Righetti, J., & Sung, L. (2000). Preventing dog bites in children: Randomised controlled trial of an educational intervention. *British Medical Journal, 320,* 1512–1513.

Chiandetti, C., Avella, S., Fongaro, E., & Cerri, F. (2016). Can clicker training facilitate conditioning in dogs? *Applied Animal Behaviour Science, 184,* 109–116.

Clarke, N. M., & Fraser, D. (2013). Animal control measures and their relationship to the reported incidence of dog bites in urban Canadian municipalities. *Canadian Veterinary Journal, 54,* 145–149.

Cooper, J. J., Cracknell, N., Hardiman, J., Wright, H., & Mills, D. (2014). The welfare consequences and efficacy of training pet dogs with remote electronic training collars in comparison to reward based training. *PLoS One, 9,* e102722.

Cottam, N., & Dodman, N. H. (2009). Comparison of the effectiveness of a purported anti-static cape (the Storm Defender* vs. a placebo cape in the treatment of canine thunderstorm phobia as assessed by owners' reports. *Applied Animal Behaviour Science, 119,* 78–84.

Cottam, N., Dodman, N. H., & Ha, J. C. (2013). The effectiveness of the Anxiety Wrap in the treatment of canine thunderstorm phobia: An open-label trial. *Journal of Veterinary Behavior, 8,* 154–161.

Cracknell, N. R., & Mills, D. S. (2008). A double-blind placebo-controlled study into the efficacy of a homeopathic remedy for fear of firework noises in the dog (Canis familiaris). *The Veterinary Journal, 177,* 80–88.

Cronin, G. M., Hemsworth, P. H., Barnett J. L., Jongman, E. C., Newman, E. A., & McCauley, I. (2003). An anti-barking muzzle for dogs and its short-term effects on behaviour and saliva cortisol concentrationsApplied Animal Behaviour Science, *83,* 215–226.

Demant, H., Ladewig, J., Balsby, T. J. S., & Dabelsteen, T. (2011). The effect of frequency and duration of training sessions on acquisition and long-term memory in dogs. *Applied Animal Behaviour Science, 133,* 228–234.

Department for Environment, Food and Rural Affairs. (2007). *The Dangerous Wild Animals Act 1976.* London: Department for Environment, Food and Rural Affairs.

Department for Environment, Food and Rural Affairs. (2009). *Dangerous Dogs Law: Guidance for enforcers.* London: Department for Environment, Food and Rural Affairs.

Department for Environment, Food and Rural Affairs. (2018). *Electronic training collars for cats and dogs in England. Summary of responses and government response.* London: Department for Environment, Food and Rural Affairs.

Demirbas, Y. S., Ozturk, H., Emre, B., Kockaya, M., Ozvardar, T., & Scott, A. (2016). Adults' ability to interpret canine body language during a dog–child interaction. *Anthrozoös, 29,* 581–596.

DePorter, T. L., Bledsoe, B. L., Beck, A., & Ollivier, E. (2019). Evaluation of the efficacy of an appeasing pheromone diffuser product vs placebo for management of feline aggression in multi-cat households: A pilot study. *Journal of Feline Medicine and Surgery, 21,* 293–305.

Dietz, L., Arnold, A. K., Goerlich-Jansson, V. C., & Vinke, C. M. (2018). The importance of early life experiences for the development of behavioural disorders in domestic dogs. *Behaviour, 155,* 83–114.

Dixon, C. A., Mahabee-Gittens, E. M., Hart, K. W., & Lindsell, C. J. (2012). Dog bite prevention: An assessment of child knowledge. *Journal of Pediatrics, 160,* 337–341.

Dodge, K. A. (1986). A social information processing model of social competence in children. In M. Perlmutter (Ed.), *Minnesota symposium on child psychology* (Vol. 18, pp. 77–125). Hillsdale, NJ: Eribaum.

Dorey, N. R., & Cox, D. J. (2018). Function matters: A review of terminological differences in applied and basic clicker training research. *PeerJ, 6,* e5621.

Duperrex, O., Blackhall, K., Burri, M., & Jeannot, E. (2009). Education of children and adolescents for the prevention of dog bite injuries (Review): Issue 2. *The Cochrane Database of Systematic Reviews.*

Eysenck, H. J., & Eysenck, S. B. G. (1964). *Manual of the Eysenck Personality Questionnaire.* London: University of London Press.

Eysenck, S. B. G., & Eysenck, H. J. (1972). The questionnaire measurement of psychoticism. *Psychological Medicine, 2,* 50–55.

Farhoody. P., Mallawaarachchi, I., Tarwater, P. M., Serpell, J. A., Duffy, D. L., & Zink, C. (2018). Aggression toward familiar people, strangers, and conspecifics in gonadectomized and intact dogs. *Frontiers in Veterinary Science, 5,* Article 18.

Feng, L. C., Howell, T. J., & Bennett, P. C. (2016). How clicker training works: Comparing reinforcing, marking, and bridging hypotheses. *Applied Animal Behaviour Science, 181,* 34–40.

Feng, L. C., Howell, T. J., & Bennett, P. C. (2018). Practices and perceptions of clicker use in dog training: A survey-based investigation of dog owners and industry professionals. *Journal of Veterinary Behavior, 23,* 1–9.

Fernandes, J. G., Olsson, I. A. S., & de Castro, A. C. V. (2017). Do aversive-based training methods actually compromise dog welfare? A literature review. *Applied Animal Behaviour Science, 196,* 1–12.

Firnkes, A., Bartels, A., Bidoli, E., & Erhard, M. (2017). Appeasement signals used by dogs during dog–human communication. *Journal of Veterinary Behavior, 19,* 35–44.

Fukuzawa, M., & Hayashi, N. (2013). Comparison of 3 different reinforcements of learning in dogs (*Canis familiaris*). *Journal of Veterinary Behavior, 8,* 221–224.

Fugazza, C., & Miklósi, A. (2015). Social learning in dog training: The effectiveness of the Do as I do method compared to shaping/clicker training. *Applied Animal Behaviour Science, 171,* 146–151.

Fuseini, A., Knowles, T. G., Hadley, P. J., & Wotton, S. B. (2017). Food and companion animal welfare: The Islamic perspective. *CAB Reviews, 12,* 1–6.

Gazzano, A., Zilocchi, M., Massoni, E., & Mariti, C. (2013). Dogs' features strongly affect people's feelings and behavior toward them. *Journal of Veterinary Behavior, 8,* 213–220.

Gershoff, E. T., Goodman, G. S., Miller-Perrin, C. L., Holden, G. W., Jackson, Y., & Kazdin, A. E. (2018). The strength of the causal evidence against physical punishment of children and its implications for parents, psychologists, and policymakers. *American Psychologist, 73,* 626–638.

González-Martínez, A., Martínez, M. F., Rosado, B., Luño, I., Santamarina, G., Suárez, M. L.,… Diéguez, F. J. (2019). Association between puppy classes and adulthood behavior of the dog. *Journal of Veterinary Behavior, 32,* 36–41.

Goode, E., & Ben-Yehuda, N. (1994). *Moral panics: The social construction of deviance.* Oxford: Blackwell.

Greenebaum, J. B. (2010). Training dogs and training humans: Symbolic interaction and dog training. *Anthrozoös, 23,* 129–141.

Grigg, E. K. (2019). Helping clients facing behavior problems in their companion animals. In L. Kogan & C. Blazina (Eds.), *Clinician's guide to treating companion animal issues* (pp. 281–317). Cambridge, MA: Academic Press.

Grohmann, K., Dickomeit, M. J., Schmidt, M. J., & Kramer, M. (2013). Severe brain damage after punitive training technique with a choke chain collar in a German shepherd dog. *Journal of Veterinary Behavior, 8,* 180–184.

Gruen, M. E., Thomson, A. E., Clary, G. P., Hamilton, A. K., Hudson, L. C., Meeker, R. B., & Sherman, B. L. (2013). Conditioning laboratory cats to handling and transport. *Lab Animal, 42,* 385–389.

Hallsworth, S. (2011). Then they came for the dogs! *Crime, Law, and Social Change, 55,* 391–403.

Hammerle, M., Horst, C., Levine, E., Overall, K., Radosta, L., Rafter-Ritchie, M., & Yin, S. (2015). 2015 AAHA canine and feline behavior management guidelines. *Journal of the American Animal Hospital Association, 51,* 205–221.

Hemenway, D. (2013). Three common beliefs that are impediments to injury prevention. *Injury Prevention, 19,* 290–293.

Herron, M. E., Shofer, F. S., & Reisner, I. R. (2009). Survey of the use and outcome of confrontational and non-confrontational training methods in client-owned dogs showing undesired behaviors. *Applied Animal Behaviour Science, 117,* 47–54.

Hollin, C. R. (1995). The meaning and implications of "programme integrity." In J. McGuire (Ed.), *What works: Effective methods to reduce reoffending* (pp. 195–208). Chichester, West Sussex: John Wiley & Sons.

Hollin, C. R., & Trower, P. (Eds.). (1986). *Handbook of social skills training, Volume 1: Applications across the life span*. Oxford: Pergamon Press.

Hughes, G., Maher, J., & Lawson, C. (2011). *Status dogs, young people and criminalisation: Towards a preventative strategy*. Report for The Royal Society for the Prevention of Cruelty to Animals. Cardiff Centre for Crime, Law and Justice: Cardiff University.

Hulac, D., Benson, N., Nesmith, M. C., & Shervey, S. W. (2016). Using variable interval reinforcement schedules to support students in the classroom: An introduction with illustrative examples. *Journal of Educational Research and Practice, 6*, 90–96.

Isbell, C., & Pavia, A. (2009). *Rabbits for dummies* (2nd ed.). Hoboken, NJ: John Wiley & Sons.

Johnson-Bennett, P. (2011). *Think like a cat: How to raise a well-adjusted cat—Not a sour puss*. New York: Penguin Books.

Kasbaoui, N., Cooper, J., Mills, D. S., & Burman, O. (2013). Effects of long-term exposure to an electronic containment system on the behaviour and welfare of domestic cats. *PLoS One, 11*, e0162073.

Kaufmann, C. A., Forndran, S., Stauber, C., Woerner, K., & Gansloßer, U. (2017). The social behaviour of neutered male dogs compared to intact dogs (*Canis lupus familiaris*): Video analyses, questionnaires and case studies. *Veterinary Medicine Open Journal, 2*, 22–37.

King, C., Buffington, L., Smith, T. J., & Grandin, T. (2014). The effect of a pressure wrap (ThunderShirt®) on heart rate and behavior in canines diagnosed with anxiety disorder. *Journal of Veterinary Behavior, 9*, 215–221.

Kogan, L., Kolus, C., & Schoenfeld-Tacher, R. (2017). Assessment of clicker training for shelter cats. *Animals : An Open Access Journal from MDPI, 7*, 73.

Lakestani, N. N., & Donaldson, M. L. (2015). Dog bite prevention: Effect of a short educational intervention for preschool children. *PLoS One, 10*, e0134319.

Lakestani, N. N., Donaldson, M. L., & Waran, N. (2014). Interpretation of dog behavior by children and young adults. *Anthrozoös, 27*, 65–80.

Landsberg, G. M., Beck, A., Lopez, A., Deniaud, M., Araujo, J. A., & Milgram, N. W. (2015). Dog-appeasing pheromone collars reduce sound-induced fear and anxiety in beagle dogs: A placebo-controlled study. *The Veterinary Record, 177*, 260.

Landsberg, G. M., Milgram, N. W., Mougeot, I., Kelly, S., & de Rivera, C. (2017). Therapeutic effects of an alpha-casozepine and L-tryptophan supplemented diet on fear and anxiety in the cat. *Journal of Feline Medicine and Surgery 19*, 594–602.

Laurin, K., Shariff, A. F., Henrich, J., & Kay, A. C. (2012). Outsourcing punishment to God: Beliefs in divine control reduce earthly punishment. *Proceedings of the Royal Society B: Biological Sciences, 279*, 3272–3281.

Lees, P., Chambers, D., Pelligand, L., Toutain, P-L., Whiting, M., & Whitehead. M. L. (2017a). Comparison of veterinary drugs and veterinary homeopathy: Part 1. *The Veterinary Record, 181*, 170–176.

Lees, P., Pelligand, L., Whiting, M., Chambers, D., Toutain, P. L., & Whitehead, M. L. (2017b). Comparison of veterinary drugs and veterinary homeopathy: Part 2. *The Veterinary Record, 181*, 198–207.

Levine, E. D., & Mills, D. S. (2008). Long-term follow-up of the efficacy of a behavioral treatment programme for dogs with firework fears. *The Veterinary Record, 162*, 657–659.

Levine, E. D., Ramos, D., & Mills, D. S. (2007). A prospective study of two self-help CD based desensitization and counter-conditioning programmes with the use of Dog Appeasing Pheromone for the treatment of firework fears in dogs (*Canis familiaris*). *Applied Animal Behaviour Science, 105*, 311–329.

Lie, M. S. B. (2017). "Stepdogs" of society: The impact of breed bans in Norway. *Critical Criminology, 25*, 293–309.

Lockhart, J., Wilson, K., & Lanman, C. (2013). The effects of operant training on blood collection for domestic cats. *Applied Animal Behaviour Science, 143*, 128–134.

Maher, J., & Pierpoint, H. (2011). Friends, status symbols and weapons: The use of dogs by youth groups and youth gangs. *Crime, Law, and Social Change, 55*, 405–420.

Makowska, I. J. (2018). *Review of dog training methods: Welfare, learning ability, and current standards.* Vancouver, Canada: BC SPCA.

Mariti, C., Ciceroni, C., & Sighieri, C. (2015). Italian breed-specific legislation on potentially dangerous dogs (2003): Assessment of its effects in the city of Florence (Italy). *Dog Behavior, 1*, 25–31.

Mariti, C., Falaschi, C., Zilocchi, M., Fatjó, J., Sighieri, C., Ogi, A., & Gazzano, A. (2017). Analysis of the intraspecific visual communication in the domestic dog (Canis familiaris): A pilot study on the case of calming signals. *Journal of Veterinary Behavior, 18*, 49–55.

Masson, S., de la Vega, S., Gazzano, A., Mariti, C., Da Graça Pereira, G., Halsberghe, C.,… & Schoening, B. (2018). Electronic training devices: Discussion on the pros and cons of their use in dogs as a basis for the position statement of the European Society of Veterinary Clinical Ethology. *Journal of Veterinary Behavior, 25*, 71–75.

Masson, S., Nigron, I., & Gaultier, E. (2018). Questionnaire survey on the use of different e-collar types in France in everyday life with a view to providing recommendations for possible future regulations. *Journal of Veterinary Behavior, 26*, 48–60.

Mathie, R. T., & Clausen, J. (2015a). Veterinary homeopathy: Systematic review of medical conditions studied by randomised trials controlled by other than placebo. *BMC Veterinary Research, 11*, 236.

Mathie, R. T., & Clausen, J. (2015b). Veterinary homeopathy: Meta-analysis of randomised placebo-controlled trials. *Homeopathy, 104*, 3–8.

McCarthy, D. (2016). Dangerous dogs, dangerous owners and the waste management of an 'irredeemable species'. *Sociology, 50*, 560–575.

Meints, K., Brelsford, V., & De Keuster, T. (2018). Teaching children and parents to understand dog signaling. *Frontiers in Veterinary Science, 5*, Article 257.

Meints, K., & de Keuster, T. (2009). Brief Report: Don't kiss a sleeping dog: The first assessment of "The Blue Dog" bite prevention program. *Journal of Pediatric Psychology, 34*, 1084–1090.

Menache, S. (1997). Dogs: God's worst enemies? *Society and Animals, 5*, 23–44.

Michelazzi, M., Berteselli, G. V., Talamonti, Z., Cannas, S., Scaglia, E., Minero, M., & Palestrini, C. (2015). Efficacy of L-Theanine in the treatment of noise phobias in dogs: Preliminary results. *Journal of Veterinary Behavior Clinical Applications and Research, 29*, 53–59.

Mills, D. S. (2005). What's in a word? A review of the attributes of a command affecting the performance of pet dogs. *Anthrozoös, 18*, 208–221.

Mongillo, P., Adamelli, S., Pitteri, E., & Marinelli, L. (2015). Attention of dogs and owners in urban contexts: Public perception and problematic behaviors. *Journal of Veterinary Behavior, 10*, 210–216.

Morrongiello, B. A., Schwebel, D. C., Stewart, J., Bell, M., Davis, A. L., & Corbett, M. R. (2013). Examining parents' behaviors and supervision of their children in the presence of an unfamiliar dog: Does The Blue Dog intervention improve parent practices? *Accident Analysis and Prevention, 54*, 108–113.

Nilson, F., Damsager, J., Lauritsen, J., & Bonander, C. (2018). The effect of breed-specific dog legislation on hospital treated dog bites in Odense, Denmark—A time series intervention study. *PLoS One, 13*, e0208393.

Olson, K. R., Levy, J. K., Norby, B., Crandall, M. M., Broadhurst, J. E., Jacks, S., Barton, R. C., & Zimmerman, M. S. (2015). Inconsistent identification of pit bull-type dogs by shelter staff. *Veterinary Journal, 206*, 197–202.

Orritt, R., Gross, H., & Hogue, T. (2015). His bark is worse than his bite: Perceptions and rationalization of canine aggressive behavior. *Human-Animal Interaction Bulletin, 3*, 1–20.

Ott, S. A., Schalke, E., von Gaertner, A. M., & Hackbarth, H. (2008). Is there a difference? Comparison of golden retrievers and dogs affected by breed-specific legislation regarding aggressive behavior. *Journal of Veterinary Behavior, 3*, 134–140.

Overall, K. L., & Dunham, A. E. (2002). Clinical features and outcome in dogs and cats with obsessive-compulsive disorder: 126 cases (1989–2000). *Journal of the American Veterinary Medical Association, 221*, 1445–1452.

Oxley, J. A., Christley, R., & Westgarth, C. (2018). Contexts and consequences of dog bite incidents. *Journal of Veterinary Behavior, 23*, 33–39.

Pageat, P., & Gaultier, E. (2003). Current research in canine and feline pheromones. *The Veterinary Clinics Small Animal Practice, 33*, 187–211

Parliament of Victoria. (2016). *Inquiry into the legislative and regulatory framework relating to restricted-breed dogs*. Victoria, Australia: Parliament of Victoria, Legislative Council Economy and Infrastructure Committee.

Pekkin, A-M., Hänninen, L., Tiirad, K., Koskela, A., Pöytäkangas, M., Lohic, H., & Valros, A. (2016). The effect of a pressure vest on the behaviour, salivary cortisol andurine oxytocin of noise phobic dogs in a controlled test. *Applied Animal Behaviour Science, 185*, 86–94.

Pineda, S., Anzola, B., Olivares, A., & Ibáñez, M. (2014). Fluoxetine combined with clorazepate dipotassium and behaviour modification for treatment of anxiety-related disorders in dogs. *Veterinary Journal, 199*, 387–391.

Pratsch, L., Mohra, N., Palme, R., Rost, J., Troxler, J., & Arhant, C. (2018). Carrier training cats reduces stress on transport to a veterinary practice. *Applied Animal Behaviour Science, 206*, 64–74.

Protopopova, A., Kisten, D., & Wynne, C. (2016). Evaluating a humane alternative to the bark collar: Automated differential reinforcement of not barking in a home-alone setting. *Journal of Applied Behavior Analysis, 49*, 735–744.

Pryor, K. (1999). *Don't shoot the dog! The new art of teaching and training* (2nd ed.). New York: Bantam Books.

Radosta, L. (2016, July). Storm phobia in dogs. *Veterinary Team Brief*, 33–37.

Reese, J. F. (2005). Dogs and dog control in developing countries. In D. J. Salem & A. N. Rowan (Eds.), *The state of the animals III: 2005* (pp. 55–64). Washington, DC: Humane Society Press.

Rezac, P., Rezac, K., & Slama, P. (2015). Human behavior preceding dog bites to the face. *Veterinary Journal, 206*, 284–288.

Riemer, S., Ellis, S. L. H., Thompson, H., & Burman, O. H. P. (2018). Reinforcer effectiveness in dogs—The influence of quantity and quality. *Applied Animal Behaviour Science, 206*, 87–93.

Rooney, N. J., & Cowan, S. (2011). Training methods and owner–dog interactions: Links with dog behaviour and learning ability. *Applied Animal Behaviour Science, 132*, 169–177.

Royal Society for the Prevention of Cruelty to Animals. (2010). *Briefing note on dangerous dogs*. Southwater, West Sussex: RSPCA.

Sarenbo, S. L. (2019). Canines seized by the Swedish Police Authority in 2015–2016. *Forensic Science International, 296*, 101–109.

Sargisson, R. J. (2014). Canine separation anxiety: Strategies for treatment and management. *Veterinary Medicine: Research and Reports, 5*, 143–151.

Schalke, E., Ott, S. A., von Gaertner, A. M., Hackbarth, H., & Mittmann, A. (2008). Is breed-specific legislation justified? Study of the results of the temperament test of Lower Saxony. *Journal of Veterinary Behavior, 3*, 97–103.

Schilder, M. B. H., & van der Borg, J. A. M. (2004). Training dogs with help of the shock collar: Short and long term behavioural effects. *Applied Animal Behaviour Science, 85*, 319–334.

Schwebel, D. C., Morrongiello, B. A., Davis, A. L., Stewart, J., & Bell, M. (2012). *The Blue Dog*: Evaluation of an interactive software program to teach young children how to interact safely with dogs. *Journal of Pediatric Psychology, 37*, 272–281.

Seksel, K., & Lindeman, M. J. (1998). Use of clomipramine in the treatment of anxiety-related and obsessive-compulsive disorders in cats. *Australian Veterinary Journal, 76*, 317–321.

Seksel, K., & Lindeman, M. J. (2001). Use of clomipramine in treatment of obsessive compulsive disorder, separation anxiety and noise phobia in dogs: A preliminary, clinical study. *Australian Veterinary Journal, 79*, 252–256.

Serpell, J. A., & Hsu, Y. A. (2005). Effects of breed, sex, and neuter status on trainability in dogs. *Anthrozoös, 18*, 196–207.

Shen, J., Pang, S., & Schwebel, D. C. (2016). A randomized trial evaluating child dog-bite prevention in rural China through video-based testimonials. *Health Psychology, 35*, 454–464.

Shen, J., Rouse, J., Godbole, M., Wells, H. L., Boppana, S., & Schwebel, D. C. (2017). Interventions to educate children about dog safety and prevent pediatric dog-bite injuries: A meta-analytic review. *Journal of Pediatric Psychology, 42*, 779–791.

Sheppard, G., & Mills D. S. (2003). Evaluation of dog-appeasing pheromone as a potential treatment for dogs fearful of fireworks. *The Veterinary Record, 152*, 432–436.

Siettou, C. (2019). Evaluating the recently imposed English compulsory dog microchipping policy. Evidence from an English Local Authority. *Preventive Veterinary Medicine, 163*, 31–36.

Siracusa, C. (2016). Status-related aggression, resource guarding, and fear-related aggression in 2 female mixed breed dogs. *Journal of Veterinary Behavior, 12*, 85–91.

Smith, S. M., & Davis, E. S. (2008). Clicker increases resistance to extinction but does not decrease training time of a simple operant task in domestic dogs (*Canis familiaris*). *Applied Animal Behaviour Science, 110*, 318–329.

Tibbetts, E. A. (2013). The function, development, and evolutionary stability of conventional signals of fighting ability. *Advances in the Study of Behavior, 45*, 49–80.

Todd, Z. (2018). Barriers to the adoption of humane dog training methods. *Journal of Veterinary Behavior, 25*, 28–34.

Torcivia, C., & McDonnell, S. M. (2018). Case series report: Systematic rehabilitation of specific health care procedure aversions in 5 ponies. *Journal of Veterinary Behavior, 25*, 41–51.

Turcsán, B., Range, F., Virányi, Z., Miklósi, A., & Kubinyi, E. (2012). Birds of a feather flock together? Perceived personality matching in owner–dog dyads. *Applied Animal Behaviour Science, 140*, 154–160.

van der Borg, J. A. M., Graat, E. A. M., & Beerda, B. (2017). Behavioural testing based breeding policy reduces the prevalence of fear and aggression related behaviour in Rottweilers. *Applied Animal Behaviour Science, 195*, 80–86.

Vaterlaws-Whiteside, H., & Hartmann, A. (2017). Improving puppy behavior using a new standardized socialization program. *Applied Animal Behaviour Science, 197*, 55–61.

Vertalka, J., Reese, L. A., Wilkins, M. J., & Pizarro, J. M. (2018). Environmental correlates of urban dog bites: A spatial analysis. *Journal of Urban Affairs, 40*, 311–328.

Vicars, S. M., Miguel, C. F., & Sobie, J. L. (2014). Assessing preference and reinforcer effectiveness in dogs. *Behavioural Processes, 103*, 75–83.

Vieira de Castro, A. C., Barrett, J., de Sousac, L., & Olsson, I. A. S. (2019). Carrots versus sticks: The relationship between training methods and dog-owner attachment. *Applied Animal Behaviour Science, 219*, Article 104831.

VKM Mejdell, C. M., Basic, D., & Bøe, K. E. (2017). *A review on the use of electric devices to modify animal behaviour and the impact on animal welfare. Opinion of the Panel on Animal Health and Welfare of the Norwegian Scientific Committee for Food and Environment.* VKM Report 2017:31, Norwegian Scientific Committee for Food and Environment (VKM), Oslo, Norway.

Wan, M., Bolger, N., & Champagne, F. A. (2012). Human perception of fear in dogs varies according to experience with dogs. *PLoS One, 7*, e51775.

Wells, D. L., & Hepper, P. G. (2012). The personality of "aggressive" and "non-aggressive" dog owners. *Personality and Individual Differences, 53*, 770–773.

Westgarth, C., & Watkins, F. (2015). A qualitative investigation of the perceptions of female dog-bite victims and implications for the prevention of dog bites. *Journal of Veterinary Behavior, 10*, 479–488.

Whittaker, S. (2005). *Think PET!* Rothley, Leicestershire: WitsEnd.

Wolpe, J. (1958). *Psychotherapy by reciprocal inhibition.* Stanford, CA: Stanford University Press.

Wolpe, J. (1962). Isolation of a conditioning procedure as the crucial psychotherapeutic factor: A case study. *Journal of Nervous and Mental Disease, 134*, 316–329.

Wong, P. W., & Cheong, S. K. (2010). Behaviour problems in dogs: The relationship between reported behaviour problems, causal attributions, and ineffective discipline. *Sunway Academic Journal, 7*, 16–32.

Yin, S., Fernandez, E. J., Pagan, S., Richardson, S. L., & Snyder, G. (2008). Efficacy of a remote-controlled, positive-reinforcement, dog-training system for modifying problem behaviors exhibited when people arrive at the door. *Applied Animal Behaviour Science, 113*, 123–138.

Ziv, G. (2017). The effects of using aversive training methods in dogs: A review. *Journal of Veterinary Behavior, 19*, 50–60.

Zwida, K., & Kutzler, M. A. (2016). Non-reproductive long-term health complications of gonad removal in dogs as well as possible causal relationships with post-gonadectomy elevated luteinizing hormone (LH) concentrations. *Journal of Etiology and Animal Health, 1*, 002.

Part III
Humans and animals
Friend or foe?

6 Animals amusing and assisting humans

Animals as entertainment

Entertainment in its many forms fulfils several psychological needs: for example, it can manipulate our emotions so we feel the pleasure of laughter; it can make us feel sad or angry or even depressed; it can teach us about new topics; and it can be a social event. There are familiar animals in literature: for example, Robert Louis Stevenson's 1883 novel *Treasure Island* introduced us to Long John Silver's parrot Captain Flint and its squawks of "Stand by to go about" and "Pieces of Eight." Michael Morpurgo's 1982 novel *War Horse*, later a stage play and a film, tells of young Albert, who in 1914 enlists to fight World War I after his beloved farm horse Joey is purchased for war service. Albert's search for Joey takes him to the front lines in France and the tale unfolds. The topic of animals in war is covered later in this chapter.

In film and television, there is alongside real animals a long history of puppet and cartoon animals. In some instances, the animals are the stars, as with puppets *Sooty and Sweep* or cartoon characters *Tom and Jerry*; or the animals may play supporting roles as with the Golden Retriever Nigel in the TV programme *Gardener's World* (Don, 2016), the dog Snowy in *Hergé's Adventures of TinTin*, or the horses Trigger and Silver who (literally) supported Roy Rogers and The Lone Ranger. Some animals became film stars in their own right: the dog called Lassie, a Rough Collie, starred in film and television over several decades beginning in the 1943 when Elizabeth Taylor co-starred in *Lassie Come Home*. In a long-running series, particularly when the human actors change so as to appear not to grow older, it is unlikely that an animal's lifespan will allow it to stay the course. Thus, Pal (1940–1958) was the first of 10 generations of dogs to play Lassie.

There was multi-occupancy of an animal role with Tarzan's sidekick the chimpanzee *Cheetah*, who debuted alongside Johnny Weissmuller in the 1932 film *Tarzan the Ape Man*. For most people, one chimpanzee is pretty much like another; not only did different chimpanzees play the role in different films but several chimpanzees played the part of Cheetah in the same film, depending on what activity was required. Over the series of *Tarzan* films, about 20 chimpanzees played the role.

It is not difficult to see the appeal, certainly to a western audience, of chimpanzees and they often appear in advertising. However, there may be a price to pay for the use of primates in films and advertising as their portrayal as human-like in manufactured situations may lead to public misunderstanding of their endangered status and foster a view of them as appropriate companion animals (Aldrich, 2018; Schroepfer, Rosati, Chartrand, & Hare, 2011).

Scanes (2018) includes films and advertising in a long list of the ways we use animals for our entertainment including animal parks, aquaria, game reserves, and zoos; events such as circuses, dog shows, and rodeos; riding animals for pleasure or in competition; hunting, shooting, and fishing; cultural sports such as bull fighting and illegal (in some countries) activities such as dog fighting or bear baiting.

The activities in Scanes' list appeal to two basic contingencies from which we humans gain very different psychological experiences: in the first there is, for want of a better word, an innocent use of animals to amuse or give us cause to marvel and admire; the second, altogether darker, is to find pleasure either in watching animals harm each other or in inflicting harm or death upon an animal. Neither of these contingencies is without attendant issues regarding the rights and welfare of the animals. The more moderate types of activities are discussed here; the explicitly harmful maltreatment of animals is covered in the following chapter.

Watching the animals

There are two venues where large numbers of people traditionally gathered to watch animals: the circus and the zoo.

The circus

In civilisations past, the circus was a venue for public entertainment with spectacles such as chariot races and gladiatorial combat. The ancient Greeks sought entertainment at the Hippodrome – *hippos* (horse) and *dromos* (course); citizens of the Roman Empire flocked to the Circus (Latin for *circle*) to watch the spectacle. As centuries passed, so the circus evolved into an ensemble of animals and human entertainers. In some cultures, the circus favoured human performers such as trapeze artists and acrobats (Baston, 2018) while elsewhere wild animals and displays of horsemanship in trick riding, all interspersed with clowns, were the main attraction (Lavers, 2015). The appeal of wild animals lay partly in their novelty value. In the days before television, many people would never have seen an elephant or a tiger so that the circus and the zoo presented a rare opportunity to be amazed.

There are two types of circuses: the static circus and the travelling circus. The Blackpool Tower Circus in the UK is an example of a static circus. In the travelling circus, animals are held in "beast wagons," transport containers carried on a long trailer (Iossa, Soulsbury, & Harris, 2009). There are several famous travelling circuses in the UK, such as Bertram Mills Circus and Smart's Circus; further afield, there is the Moscow Circus and the Barnum & Bailey Circus in America. While travelling long distances may be tolerable for smaller animals, it is altogether less than satisfactory for larger animals such as elephants and large cats. As Tait and Farrell (2010) discuss, in the 1970s growing popular concern with the treatment of animals led to co-ordinated protests against circuses by organisations such as *Animal Liberation* and later *People for the Ethical Treatment of Animals (PETA)*. The website *StopCircusSuffering.com* lists over 20 countries that have banned the use of wild animals in circuses. The *New Circus*, sometimes called *Nouveau Cirque*, is a more recent development which eschews animals and focuses on storytelling alongside innovative use of costume design, lighting, and music.

The zoo

While watching the animals is the obvious reward for the entrance fee, there are many reasons for going to the zoo such as a social outing, family enjoyment, education, and entertainment. In addition, many modern-day zoos perceive that as well as entertaining the public their role encompasses conservation, education, and research (Figure 6.1).

There are four aspects of zoos of particular interest: (i) the animals on show, (ii) the visitors to the zoo, (iii) programmes to educate visitors about the animals, and (iv) the impact of zoos on conservation.

Figure 6.1 Watching the animals.
Source: Photograph by Shawn Reza. Microsoft Pexels, Free Stock Photos.

Animals in captivity

All types of animals are held in zoos: the familiar vertebrates, mammals, birds, fish, reptiles and amphibians, and the sometimes less familiar invertebrates such as arthropods, cephalopods, and insects. It is impossible for any one zoo to house every animal and so most zoos will seek to display the animals that will appeal to the majority of their visitors. These "charismatic" animals are likely to be mammals such as gorillas, monkeys, elephants, giraffes, and big cats, while freshwater and marine life such as sharks and manta rays may be showcased in aquaria. In some cases, zoos may seek to hold collections of a particular type of animal as an aid to conservation (e.g., Frynta, Šimková, Lišková, & Landová, 2013) or have breeding programmes for rare animals.

Animal welfare

It is clearly in everyone's interest – the animals, the zoo, the visitors – that the welfare of the captive animals is given prime importance. The zoo's management of the animals' welfare may be guided by the "five freedom principles": (1) freedom from hunger, thirst and malnutrition; (2) freedom from discomfort and exposure; (3) freedom from pain, injury, and disease; (4) freedom from fear and distress; (5) freedom to express normal behaviour. These principles are not perfect and do not countermand the fact that some animals are held in an artificial environment. Nonetheless the freedom principles provide a guide to good practice in promoting animal welfare (Mäekivi, 2018). As zoos and the animals they contain are dynamic entities they can change continually: it is critical therefore that animal welfare is seen as constant process and monitored accordingly (Brando & Buchanan-Smith, 2018; Wark et al., 2019). However, monitoring animal welfare is not an empty exercise, the process of inspection should if necessary lead to changes to improve matters. Thus, changes may be needed in any or all of the five freedom principles as highlighted by the inspection.

A robust inspection relies in part on a comprehensive checklist to capture information across all the necessary domains (Hitchens, Hultgren, Frössling, Emanuelson, & Keeling, 2017). The critical point is that if the inspection reveals shortcomings in animal welfare – in the UK this would be judged in terms of compliance with standards set by the Zoo Licensing Act 1981 – then there should be a means by which to ensure standards are met. Draper, Browne, and Harris (2013) cast doubts over the UK system; their analysis of 136 inspection reports on British zoos led them to conclude that: "The current system of licensing and inspection does not ensure that British zoos meet and maintain, let alone exceed, the minimum animal welfare standards" (p. 1058).

The welfare of animals in captivity can be compromised in several ways, including stressful physical conditions brought about by artificial light, pervasive smells and sounds, and incorrect temperatures; the provision of insufficient space leading to confinement-specific stressors such as restricted movement and reduced retreat space; forced proximity to humans; and living in abnormal social groups (Morgan & Tromborg, 2007). When animal welfare is compromised, the effects may become evident in various ways, one of the most obvious of which is stereotypic behaviour seen in animals on farms and in laboratories and zoos. Mason (1991) notes that: "Stereotypies are repetitive, unvarying and apparently functionless behaviour patterns typical of animals in some conditions of captivity" (p. 103). Thus, stereotypic behaviours may be seen, for example, in the animal's abnormal swinging of its head, chasing its own tail, and repetitive pacing of its enclosure (Cless & Lucas, 2017). Self-harm can be a consequence of these behaviours which can persist even when the aversive conditions are removed. An explanation for stereotypic behaviour is complex, necessarily based on the fine details of atypical physical and physiological responses to an abnormal environment (Mason, 2006). Stereotypic behaviour may be a sign that attention to the animal's welfare is required (Mason & Latham, 2004; Rose, Nash, & Riley, 2017).

Given the range of threats to animal welfare and their serious consequences, what steps can be taken to reduce risks to the animals' well-being?

Enrichment

If, as Skinner's work tells us, behaviour is intertwined with environment then or most captive animals their environment is, to varying degrees, unnatural. An unnatural environment may precipitate abnormal behaviour producing welfare issues for the animals. It is impossible in every instance to reproduce exactly the animals' natural environment: the needs of aardvarks, differ from birds, which differ from fishes and so on *ad infinitum* (Fife-Cook & Franks, 2019; Patoka, Vejtrubová, Vrabec, & Masopustová, 2018; Rose, Brereton, & Croft, 2018). The zoo animal's environment is empty of natural predators and the presence of keepers, vets, and so on mean that the captive animal may have an unnaturally long lifespan. The welfare of the aging animal and the many changes, psychical and psychological, which it brings demands new skills from those responsible for the animals' well-being (Krebs, Marrin, Phelps, Krol, & Watters, 2018).

However, it is possible to ameliorate the captive animal's living conditions (Maple & Perdue, 2013). There are several strategies, not mutually exclusive, by which to enrich the captive environment: (i) improving the design of the enclosure; (ii) more realistic feeding procedures; (iii) introducing novel objects; (iv) social enrichment; and (v) bringing in new sensory stimuli. It is anticipated that these strategies will result in the enhanced behaviour, good biological functioning, and fewer welfare problems. There is a body of research that considers the effects of these strategies across a diverse range of animals including cheetahs (Quirke & O'Riordan, 2011), tigers (Szokalski, Litchfield, & Foster, 2012), elephants (Greco et al., 2016), geckos (Bashaw, Gibson, Schowe, & Kucher, 2016), and walruses (Kastelein, Jennings, & Postma, 2007).

As zoo animals will have some level of contact with people, is it possible to use these meetings as a form of enrichment? For many animals, their most frequent and immediate human contact is with their zookeepers.

Zookeepers

Hosey and Melfi (2012) asked those working in zoos about their relationships with the animals in their care. A high proportion of respondents said that they had established bonds with a zoo animal, most frequently with primates and carnivores, which brought the benefits of making the animal calm and less stressed so that it was easier to handle and give treatments. The keepers thought that the bond bestowed mutual benefits such that the animals enjoyed the contact and the enjoyment of being with the animal adding to the job satisfaction. There is evidence to support the view that the formation of keeper-animal dyads has beneficial effect on animal welfare (Ward & Melfi, 2015).

Cole and Fraser (2018) make similar points to Hosey and Melfi in pointing to the potential of the animal–human bond to enhance the welfare of zoo animals. They draw

Table 6.1 "Human Dimension" principles in good animal–human relationships (after Cole & Fraser, 2018)

1. Positive human–animal interactions, avoiding punishment.
2. Maintaining the same keepers with the same animals.
3. Taking account of individual differences between animals.
4. The keeper's attitude and personality.
5. The knowledge and experience keepers gain over time.
6. The physical and mental well-being of zoo staff.
7. The design of the physical environment to facilitate positive interactions with the animals.

on research with farm animals, where there is a commercial benefit associated with good animal–human relationships, to derive seven principles inherent in the "human dimension" of animal welfare (see Table 6.1).

The fourth point in Table 6.1 is interesting from a psychological perspective. Cole and Fraser cite Seabrook's research with farm animals which: "Identified personality traits of herdspersons on dairy farms with high milk yield. He characterized the ideal herdsperson for this species as a "confident introvert" who is also considerate, patient, independent, persevering, not meek, not talkative, and unsociable" (p. 54). While a "confident introvert" is not necessarily a specification for a social standing among people, it seems to be just fine with animals; for example, these traits were associated with the willingness of cows to enter the milking facility and reduced agitation during the milking.

Phillips and Peck (2007) found that a keeper's personality influences their behaviour towards the tigers in their care. Thus, they note that angry keepers gave fewer pats, the more conscientious keepers spent less time in play with the tigers, and the more neurotic keepers had fewer interactions with the tigers. However, the keepers' patterns of behaviour were not influenced by individual differences between the tigers.

When interacting hands-on with animals, particularly the big cats and elephants, zookeepers are putting themselves at risk of injury or worse. The alternatives are protected contact, where the keeper does not enter the animals' enclosure so not sharing space with the animals but can still touch the animals, or no physical contact. Szokalski, Litchfield, and Foster (2013) surveyed 86 zookeepers from different countries regarding their views on contact. The survey revealed that protected contact, particularly with big cats, was most frequently used handling method as it is safe and allows a close bond to develop between animal and keeper. The keepers expressed safety concerns about hands-on approaches both for themselves as well as having reservations about the poor modelling close contact provides to visitors.

The relationship between keepers and animals is clearly an important aspect of enrichment of the animals' environment. While the contact is more fleeting, what of the other humans, those who come to watch the animals, on the animals' lives?

Visitors to the zoo

The visitor effect

It is obvious that visitors will have an effect on some animals. The pressing issue is whether a "visitor effect" is positive, neutral, or negative with regard to the animals' welfare (Davey, 2007; Hosey, 2000; Sherwen & Hemsworth, 2019). It is possible to find examples of each type of effect. Cook and Hosey (1995) reported how chimpanzees and visitors engaged in interactive sequences, using both gesture and vocalisation to communicate. Cook and Hosey suggest that as well as the social benefits of these interactions, they may also lead to visitors giving food which for the animals is a highly positive outcome! Choo, Todd, and Li (2011) also found that orangutans were motivated to interact with visitors when food was a possibility.

For a minority of animals, the effects of visitors are neutral: Sherwen, Magrath, Butler, Phillips, and Hemsworth (2014) looked at the behaviour of meerkats and found that they quickly became accustomed to the presence of humans. In a study comparing sea lions and harbour seals, de Vere (2018) found that visitors "Significantly affected the behavior of resident harbor seals, but not California sea lions. The effects of visitors did not appear to be negative, and visitor ability to provision the animals with food treats may mean that any impacts could potentially be positive" (p. 169). A negative visitor effect can have serious consequences for the animals including disturbed behaviour patterns, excessive aggression, and fear (e.g., Pedersen, Sorensen, Lupo, & Marx, 2019; Schultz & Young, 2018; Sherwen et al., 2015). In rare cases, a negative effect can lead to an attack on visitors; Hosey and Melfi (2015) note that such attacks are typically caused by a visitor approaching too close or even gaining access to the enclosure, or if an animal escapes from its enclosure and acts aggressively.

However, the effect of zoo visitors is not felt equally by all animals. Queiroz and Young (2018) suggest that a negative visitor effect is most likely to be found in animals such as deer, which naturally live in open conditions and are diurnal and herbivorous. On the other hand, animals from closed habitats, such as chimpanzees, which are omnivorous and diurnal, appear to be least affected by visitors.

As well as visitors, other environmental factors such as the weather and the time of day may influence the animals' behaviour. Goodenough, McDonald, Moody, and Wheeler (2019) suggest that these other influences may well exert more influence on animal behaviour than the presence of visitors. It is easy to see how the immediacy of the effect of the zoo visitor is given prominence over more mundane influences.

Psychological experiences in the zoo

Myers, Saunders, and Birjulin (2004) explored our emotional reactions to watching animals in a zoo. It is axiomatic that different animals will elicit different emotional reactions: just think cute, fluffy puppy versus slithering, squirming snake! Myers, Saunders, and Birjulin monitored visitors' reactions to three animals: the familiar gorilla, the less well-known okapi, and snakes of which we are all aware. While the gorilla and okapi elicited similar emotions, they differed from snakes; compared to the snake, the gorillas and okapi were, for example, held to instil a sense of beauty, peace, and attraction, while the snakes rated higher for fear and disgust. This point is reinforced by Janovcová et al. (2019), who found that people had a general dislike of reptiles, although some snakes were perceived as things of beauty.

The aquarium house is a perennial favourite at the zoo, attracting large numbers of visitors. The types of animal of display, both freshwater and marine, may vary from country to country reflecting species native to the area. There is a dazzling array of aquatic possibilities: there are on display invertebrates ranging from simple sponges to more complex cnidarians such as jellyfish and sea anemone; echinoderms such as starfish, sea cucumbers, and sea urchins; and crustacea including shellfish, crabs, crayfish, and lobsters. The list goes on: there are reptiles including water snakes, marine iguanas, and turtles; cephalopods such as squid and octopus; a myriad of species of fish; and cetaceans, typically the porpoise, whale, and dolphin. However, the practice of keeping cetaceans in captivity does not meet with universal public approval. Naylor and Parsons (2019) conducted an online survey on attitudes to holding cetaceans in captivity and found that the majority of the 858 respondents were opposed to the public display of dolphins and whales. An even larger majority opposed taking free-ranging dolphins and whales from the wild for display in captivity. The inter-relationship of attitudes is illustrated by those who support keeping cetaceans in captivity also more likely to say that cetacean conservation is not important.

Some zoos have introduced touch tanks, shallow open aquaria, that to allow the visitors to touch animals such as fish, invertebrates, small sharks, and stingrays. However, while enjoyable for some visitors (arguably less so for the exhibits), it is less than certain that this tactile experience has any impact on views about the value of these animals and the need for conservation (Ogle, 2016).

One of the attractions, perhaps the main attraction, of a visit to the zoo lies in our psychological experience on encountering the animals. We may experience a range of thoughts and emotions, from amazement and fear to interest and puzzlement, particularly when we are able to get close to an active animal (Luebke, Watters, Packer, Miller, & Powell, 2016), as we observe the different types of animal. In a study conducted at Chester Zoo in England, Moss and Esson (2010) assessed visitor interest in 40 species. The most popular animals were the chimpanzees, the giraffes, and the jaguars; the least popular were the hornbills, the zebra finches, and the penguins. They conclude that their findings can be applied to visitor education in that: "There is a greater potential for learning at those species that visitors are most interested in and therefore spend most time watching" (p. 727).

Yet further, as well as different reactions to different types of animals, we may also experience varied emotional reactions to different members of the same class of animal. In a Czech study by Frynta, Pelešková, Rádlová, Janovcová, and Landová (2019), respondents were shown photographs of 101 species of amphibians and asked to rate each one. They found that anurans (frogs and toads) were the most preferred and caecilians (which resemble large worms) were least preferred. Wharton, Khalil, Fyfe, and Young (2019) make the same point for marine life: "Expressive, active, charismatic animals (e.g., dolphins, sea otters, sea lions) easily elicit empathic feelings, while less expressive, less charismatic animals like sharks, salmon, or jellyfish may struggle" (p. 161) (Figures 6.2 and 6.3).

As Frynta et al. point out, understanding our reactions to animals may be important in engaging people's interest in campaigns to save globally endangered species. In this context, *empathy* is an aspect of our psychological constitution of particular importance.

Visitor empathy

The term *empathy* is generally taken to refer to the affective and cognitive reactions that allow us vicariously to understand how another person is feeling. Thus, we may feel empathic towards the grief of someone who is bereaved, or the despair of a homeless person, or the pain of a victim of an accident. There are a variety of ways in which to define and measure empathy (Neumann, Chan, Boyle, Wang, & Westbury, 2015), but one of its important features is that it can motivate us be altruistic and help those for whom we feel empathy. Does empathy for humans equate to empathy for animals? Paul (2000) found a significant correlation between a measure of empathy for people and a similar measure for animals. However, while the two measures of empathy are significantly associated, the size of the correlation is not large (+0.26),

Figure 6.2 Frogs we like.
Source: Photograph from Open Pexels.

Figure 6.3 Caecilians, not so keen.
Source: Franco Andreone Authorises the Use of the Pictures by him Published on the Website Calphotos.berkeley.edu.

leading Paul to state that the finding does "Not offer unequivocal support for the notion that human and animal-oriented empathy represent facets of a single, broadly continuous construct" (p. 199). In keeping with this conclusion, Paul reported that high levels of human-oriented empathy were related to currently having a child or children at home; while animal-oriented empathy was related to the current of pet ownership and having had childhood pets. Of course, having children and an interest in animals are two good reasons to visit a zoo.

Luebke (2018) makes point that as some zoo visitors feel empathy for the animals they observe this raises the possibility of using the zoo as a learning environment. Luebke asked visitors to an American zoo to write down details of any "extra special" experience they may have had at any particular exhibit. The visitors' experiences were then considered in association with their scores on affective and empathic questionnaires. The reported extra special experiences focused on three areas: (i) seeing an animal's behaviours such as nurturing their infants, (ii) their own reflections on nature and conservation, and (iii) their children's reactions or experiences. Those visitors who reported a special experience gave significantly higher scores on an empathy questionnaire than those who did not record anything special happening. Some visitors wrote comments indicating a deeper concern and understanding. Luebke notes that:

> Most of these comments centered on caring thoughts and feelings about animals and the environment and how connected or similar humans are to nonhumans. Other comments were about empathic concerns about the environment and the importance of conservation. All in all, these comments demonstrate that these visitors became highly engaged within the exhibits and were not merely passive observers of animals. (p. 349)

It is clear that zoos can evoke strong feelings of empathy for animals that in turn may be associated with thoughts of the environment and the need for conservation (Pfattheicher, Sassenrath, & Schindler, 2016; Young, Khalil, & Wharton, 2018). Is it possible to capitalise on this association by introducing learning programmes to encourage empathy for animals and so lead to positive behaviour towards animals?

Education at the zoo

For many people, a visit to the zoo is a rare opportunity to watch and closely engage with the animals. The interest shown by zoo visitors offers an opportunity to teach them more about the animals, their natural habitat, the present and future dangers various species may face, and the intricacies of conservation. These various topics coalesce under the rubric of *sustainability*: (i.e., sustaining the environment for the benefit of all life, flora, and fauna including humans). Yet further, Esson and Moss (2014) argue that:

> The contextual framework that guides the direction of zoo education is based on the premise that modern zoos should inspire their visitors to care about the environment and instil a sense of personal responsibility for making behaviour changes that support sustainable lifestyles. (p. 8)

This is a more than justifiable aim but it raises two issues. First, a direct psychological challenge: How to engage people's attention and convince them to change their behaviour? Second, How do you determine if the zoo's efforts in this direction are successful?

Inspiring conservation concern

As Esson and Moss make clear, most people visit the zoo primarily to be entertained, not to be inspired to undergo personal change. However, some visitors may anticipate learning and meeting the opportunity to develop their child's moral concerns about the world (Fraser, 2009). Esson and Moss (2014) describe the *Learning Together* programme developed at Chester Zoo in the UK for single-parent families. The programme consists of three 1-day visits to the zoo with a focus on flagship species so as to capture attention before introducing personal lifestyle and behaviour changes. The first visit addressed *Dangerous Beauty* and looked at how the illegal wildlife trade exploited endangered animals for profit and ways to avoid supporting this trade. The second session, *Rainforests and Us*, considered the rainforest products we use every day and the ethical choices we can exercise when shopping. Finally, the third topic looked at *Water and Life* and emphasised the valuable nature of water and how we can conserve it.

There are many other examples of similar zoo-based projects, some with a precise focus such as the dangers to wildlife posed by litter (Brown, Ham, & Hughes, 2010; Mellish, Pearson, McLeod, Tuckey, & Ryan, 2019), others with much wider emphasis taking in issues crucial to sustain life such as how we manage energy, transport, water, and waste (Gill & Warrington, 2017). While the provision of these educational projects is admirable, the crucial point is whether they are effective in the short term by increasing knowledge and in the long term by changing behaviour.

Outcome evaluation

In their evaluation of the *Learning Together* programme, Esson and Moss reported that visitors were positive about the need for affirmative action and confirmed their personal commitment to change their lifestyle. At follow-up interviews several weeks after the initial involvement, the visitors renewed their commitment to lifestyle changes. A limitation of these findings is that behaviour change was reported by the participants and not directly observed. Self-report of a commitment to modify one's behaviour is not the same as an actual change in behaviour. Esson and Moss note that while there are many similar zoo-based studies of changes in visitor knowledge and attitude, there is a distinct lack of studies that directly measured behaviour. They conclude that: "This disparity would seem to reinforce the notion of behaviour change being an elusive variable to measure" (p. 12); a sentiment with which legions of applied researchers would undoubtably agree!

In addition to problems of measuring behaviour change, there is also the issue of co-variates: a co-variate is a variable that is not controlled during data collection but which could influence the outcome. It cannot be assumed, for example, that all zoo visitors have the same levels of interest and knowledge regarding the animals so unless this disparity is controlled in the design of the study it could lead to misleading findings. In a study of visitor learning at an aquaria, Falk and Adelman (2003) managed the co-variate problem by forming groups of minimal, moderate, and extensive conservation knowledge and attitude. The use of these groups helped avoid a confound due to variations in knowledge, allowing greater confidence in its findings.

The logistical and finanical impediments to high-quality follow-up, ideally longitudinal research does not mean that evaluation should be abandoned. Collins et al. (2019) illustrate this point with a study of the effectiveness of an educational programme *within* the zoo. The programme aimed to reduce negative visitor behaviours such as feeding and shouting at the animals and banging on tanks in an aquaria. Groups of children took part in a 1-hour class that covered the biology of penguins and lemurs, the threats to their existence in their natural environment wild, and how life may differ for animals in the zoo and in the wild. The children learned, for example, that feeding animals could make them sick and that and banging on glass walls may frighten the animals. After the lesson, the children's behaviour was assessed when they observed the animals: compared to children who had not particiapted in the programme, the informed children were significantly less likely to display negative behaviours.

Learning from the evaluations

The main focus of the evaluations has been whether or not visitors learn from the available exhibits and the effectiveness of educational programmes. Pavitt and Moss (2019) contrasted traditional "stand and stare" exhibits with the more recent style of "walk-through" exhibits. They found that walk-through exhibits produced more comments than the traditional style, particularly if a volunteer or member of staff was on hand to guide discussion. In addition, vistors stayed for up to six times longer at walk-through exhibits. There were some changes in visitors' pro-conservation attitudes, but little to indicate that visitors had learned something new from the exhibit. Overall, walk-through exhibits that use staff or volunteers can enhance visitor engagement with a given species, although it is doubtful whether this translates into anything further in terms of changing visitor behaviour.

An American study by Kopczak, Kisiel, and Rowe (2015) recorded the conversations of families (with at least one adult and one child) as they engaged with the touch tanks at four zoos. They found that family discussions about ecology were limited in scope and not influenced by the quality of the display, which is at variance with studies that show tidepools with greater biodiversity generate more interest (Fairchild, Fowler, Pahl, & Griffin, 2018). However, when a volunteer or a member of staff was present for visitors to talk with, there was significantly more discourse about the exhibit. Kopczak, Kisiel, and Rowe's findings emphasise the point that visitor discussions are enhanced and potentially more beneficial when someone knowledgeable is available to encourage engagement with wider considerations.

The contemporary zoo may well be involved in conservation efforts for threatened species. The contribution made by zoos to wider conservation efforts is discussed in Chapter 9.

As well as watching the animals for entertainment, working partnerships with animals are formed in various aspects of life. As noted below, there are animals who work alongside humans in a range of activities such as guard dogs and sheep dogs. However, from a psychological perspective, the involvement of animals in therapy is of particular interest.

Working animals

Animals in the courtroom

The requirement to take part in a trial as either a witness or a victim can be a highly stressful experience, perhaps particularly so for young children. The role of a courthouse dog is to provide support for those facing psychological and emotional difficulties in taking part in proceedings (Dellinger, 2009). In the USA, the dogs are trained by Assistance Dogs International and are used in 15 states (see Courthouse Dogs at https://courthousedogs.org).

In the UK, the use of dogs in the courtroom is just emerging as a possibility (Spruin & Mozova, 2018).

Animal-assisted education

"What are all those dogs doing at school?" asks Jalongo (2005). One answer is helping children to learn to read. The reasoning behind the use of dogs to help reading is that some children may find it difficult to read aloud to a teacher and a dog provides an uncritical ear. A dog may also be easier for children to talk to – "Can you explain that big word to Rover?" – allowing them to grow in confidence in their reading skills (see Fung, 2017). A review by Hall, Gee, and Mills (2016) supported the efficacy of classroom dogs as an aid to reading, although the research is not always of the highest quality. This conclusion was broadly supported by Brelsford, Meints, Gee, and Pfeffer (2017), who, alongside dogs, added rabbits and guinea pigs as assisting with a range of classroom tasks including social functioning and interpersonal skills, adherence to instructions, and classroom behaviour. Brelsford et al. also point to the various populations, both traditional and special needs, with whom animals have been used. In keeping with Hall, Gee, and Mills, Brelsford et al., noted that the research is of variable quality.

Assistance dogs

Dogs may come to our assistance in a variety of ways. Audrestch et al. (2015) state that the three most frequently found roles are as guide dogs for visual impairment, as hearing dogs, and mobility assistance dogs. Martellucci et al. (2019) expand the range, adding autism support dogs, diabetic alert dogs, emotional support dogs, foetal alcohol spectrum disorders service dogs, psychiatric service dogs, seizure alert, and response dogs.

Sight dogs

There is nothing new about dogs assisting people with physical disabilities. Fishman (2003) notes that guide dogs for the blind were in use in 13th-century Europe and China through to the 1920s when shepherd dogs were being trained to lead blind German veterans of World War I. In the UK, the *Guide Dogs for the Blind Association* was established in 1934 which, among other functions, contributes to the training and supply of guide dogs; *The Guide Dogs of America* and *The Seeing Eye* are similar organisations in the USA.

A sight dog plays an important role in its owner's life and so it is crucial that dogs are selected that have an aptitude for this task, are physically fit, and are amenable to training. In selecting a dog for training as a sight dog, there are desirable behavioural traits to screen-in, such as a being quick to learn new tasks, and undesirable traits, such as a fear of stairs, to screen-out (Serpell & Hsu, 2001). These traits are partly a function of breed, with Golden Retrievers and Labradors often selected, and partly a function of the dog's experiences as a puppy (Batt, Batt, Baguley, & McGreevy, 2010). It is more efficient in terms of time and money to select dogs for training that are likely to be successful. A Belgian study by Bogaerts et al. (2019) estimated the cost of 40 percent drop-out rate from guide dog training programmes as €10,524 per dog. Given the importance of selecting the right dogs, several studies have considered the predictors of successful training, such as temperament (Tomkins, Thomson, & McGreevy, 2011); the dog's degree of laterization, the ability to use both paws (Batt, Batt, Baguley, & McGreevy, 2008); and formal measures of juvenile behaviour, such as jumping and barking, which may improve selection for training (Harvey et al., 2016).

Hearing dogs

We attend to a variety of sounds, verbal and non-verbal, so the hearing dog must be able to alert their owner to a range of sounds. Non-verbal sounds include a knock at the door or a ring off the doorbell, a smoke or fire alarm, a kitchen timer, an alarm clock, the telephone, and traffic noise; the verbal sounds include presence of another person, a child, and one's name being called (see Rintala, Matamoros, & Seitz, 2008; Martellucci et al., 2019). When alerted by the touch of their dog, the owner can ask, verbally or by a signal, about the source of the noise. The dog then leads them to the cause of the noise or may lie prone if there is a fire alarm.

Mobility dogs

Audrestch et al. (2015) note that mobility assistance dogs are sometimes classified as *service dogs*, although the labelling of different types of assistance dogs is a less than straightforward task (Parenti, Foreman, Meade, & Wirth, 2013). The *service dog* classification includes autism assistance dogs, medical detection dogs, and psychological assistance dogs (see below). A mobility dog is trained to provide practical assistance to a person whose impairments make some everyday tasks, such as switching on the lights or opening a door, difficult or painful to carry out.

Effects of assistance dogs

Once trained and placed with an owner, what impact do sight dogs have on the lives of those they guide? Winkle, Crowe, and Hendrix (2012) suggest three areas where positive and negative changes may be expected: these are (i) social activity, (ii) functional tasks such as performing daily chores, and (iii) psychological consequences. From the onset, the ownership of an assistance dog may require some psychological and social adjustment. An American survey by Rintala et al. (2008) looked at the experiences of people with hearing and mobility assistance dogs. Some respondents described unwanted attention from other people when in public and being confronted about taking the dog into a restaurant; others said that the dog required a great deal of attention and, as dogs do, it woke early in morning, shed fur, and showed undue interest in the garbage. As discussed in Chapter 3, there are occasions when a guide dog is attacked by another dog.

Li, Kou, Lam, Lyons, and Nguyen (2019) carried out telephone interviews with seven people aged 55 years and older with a visual impairment who were first-time dog guide owners. After a settling-in period, typically of 3 to 12 months, the new owners reflected on their responsibilities for looking after the new dog and changes in their daily habits. These new demands were balanced against bonding with the dog, an increase in community integration, and enhancement of their personal independence.

A South African study by Wiggett-Barnard and Steel (2008) involved interviewing blind people about their experience of owning a guide dog. There were eight themes, broadly in sympathy with Winkle, Crowe, and Hendrix, that emerged from the interviews: (1) the dog improves mobility, (2) the dog gives companionship, (3) the dog necessitates personal change, (4) lifestyle changes are caused by dog ownership, (5) dogs attract social attention, (6) when the dog is distracted its ability to guide is inhibited, (7) other people's ignorance about guide dogs, and (8) the dog can become a source of pride. There are two broad points within these themes: first, there is the impact of the dog as a guide; second, like any dog owner, there is the formation of an attachment with the dog as a companion. From their interviews with 63 guide dog owners, Craigon et al. (2017) found a mixture of positive and negative aspects to having a guide dog. The inevitable conclusion about a guide dog is, as they say, "She's a dog at the end of the day."

Finally, the advantages and disadvantages of owning an assistance dog are not peculiar to sight dogs. Gravrok, Bendrups, Howell, and Bennett (2019) found first-time assistance dog owners reported physical challenges such as keeping the dog under control during walks and countering the dog's distractibility. There are financial considerations to owning an assistance dog. People with hearing and mobility assistance dogs in the American survey reported by Rintala et al. (2008) said that while owning the dog they were between $175 and $6,500 out of pocket. The greatest cost came from payment of $6,500 to the organisation that trained the dog, followed by $5,500 for a dog's surgery. From a psychological perspective, the benefits of owning a guide dog range from increased self-esteem, feelings of greater independence and control, enhanced well-being and mood, and assertiveness (e.g., Audrestch et al., 2015). These positive changes may correspond with enhanced social contact and a decrease in loneliness.

Animal-assisted detection

The dog's olfactory system is physiologically very different to that of humans. The canine nose has up to 300 million olfactory receptors, compared to about 6 million in a human nose. These nasal receptors are located among turbinate bones that form numerous cylindrical passages that allow air exposure to millions more cells than is possible with a simple tubular nasal passage, as in humans. If laid out, the surface area of the dog's olfactory cells would cover the area of the dog's skin. In comparison, the surface area of human olfactory cells would just about cover a postage stamp. In addition, 40 times more of the canine brain than the human brain is given to processing the input from these receptors.

These figures indicate that the acuity of a dog's sense of small, depending on the breed, is of the order of tens of thousands of times greater than ours. This sense of smell allows the dog to detect minute amounts of an odour and to discriminate between different odours. The top five breeds for sense of smell are first, sometimes called a nose with a dog attached, the Bloodhound, followed by the Basset Hound, Beagle, German Shepherd, and Labrador Retriever. We humans are continually looking to find things that are lost or hidden, so we enrolled the dog's sense of smell. As long ago as the 17th century when St. Bernards were used to find lost travellers, we learned to put the dog's keen sense of smell to work to detect what we seek.

As with all assistance dogs, suitable dogs need to be identified and trained (as do their handlers) for specific types of detection work (e.g., DeMatteo, Davenport, & Wilson, 2019; La Toya, Baxter, & Murray, 2017; Lazarowski et al., 2018). The specialisations include explosive material detection, cargo inspection, air passenger screening, finding concealed people, and pest and agricultural pathogen detection. There are several variables that cut across these specialisations – such as urban versus wilderness search and rescue, tracking fresh and old human trails, and detecting bulk versus traces of specific substances – to consider in selection and training.

What attributes make a good search dog? Rooney, Bradshaw, and Almey (2004) surveyed representatives of the six principal UK services that use search dogs – Army (Royal Veterinary Corps and Royal Military Police), Royal Air Force, Ministry of Defence Police, HM Prison Service, HM Customs and Excise, and civilian police force – for their views on the most important characteristics of this type of dog. They reported, as shown in Table 6.2, 10 characteristics that defined a good search dog. However, it is not all about the dog: the role of the handler cannot be underestimated in the task of detection. Lit, Schweitzer, and Oberbauer (2011) conducted a study of handler-dog behaviour in a search task and concluded that: "Handler beliefs affect working dog outcomes, and human indication of scent location affects distribution of alerts more than dog interest in a particular location" (p. 393). Indeed, it is the melding of the individual functioning of the dog and the handler that brings about optimum search performance (Troisi, Mills, Wilkinson, & Zulch, 2019).

The specialist skills of detection dogs and their handlers are put to work in three principal domains: (i) the detection of humans, (ii) detection of other animals, and (iii) detection of inorganic substances.

Human detection

Various agencies use detection dogs including customs officers, medical agencies, the military services, the police, and search-and-rescue services. These agencies may use detection dogs in several ways: to track or locate suspects in a crime, to search for illegal substances such as drugs and explosives, and to detect items of contraband. While the dog's ability is paramount, the skills of the handler play a role in the outcome (Lasseter, Jacobi, Farley, & Hensel, 2003).

As Stockham, Slavin, and Kift (2004) note, scent detection may be important in criminal investigations: for example, as part of evidence gathering in cases of suspected arson, dogs can be trained to detect accelerants (Tiira, Viitala, Turunen, & Salonen, 2019). The *cadaver dog* is trained to detect human remains and alongside criminal investigations may be a valuable addition to sciences such as geophysics and soil analysis (Larson, Vass, & Wise, 2011). The cadaver dog is able to detect remains as small as an individual tooth (Cablk & Sagebiel, 2011) or as large as a mass grave (Blau et al., 2018), and remains scattered over a large surface and interfered with by

Table 6.2 Ten desirable traits in specialist search dogs (from Rooney et al., 2004)

1. Acuity of sense of smell.
2. Incentive to find an object which is out of sight.
3. Good health.
4. Tendency to hunt by smell alone.
5. Good stamina.
6. Good ability to learn from being rewarded.
7. Low distractibility when searching.
8. Good agility.
9. Overall behavioural consistency on a day-to-day basis.
10. Highly motivated to chase an object.

scavengers (Komar, 1999). While a mass grave may mark the site of a terrible crime, it may also be an ancient, forgotten cemetery of archaeological, historical, and religious interest (Baxter & Hargrave, 2015). In this light, Glavaš and Pintar (2019) find that cadaver dogs can be a valuable aid to finding archaeological sites.

A search for human remains that takes place after a disaster may take place in an environment that is gruelling and dangerous for dogs and handlers. Migala and Brown (2012) describe the use of human remains detection dogs following the wildfires in Texas in 2011 that destroyed over 1,600 homes. The dogs and their handlers searched for human remains in an extremely aversive environment with high temperatures, swirling ash that affected breathing, and the threat of new flare-ups. Despite precautions, several dogs suffered minor burns to their pads, one dog's forepaw was lacerated and required sutures, and one dog became dehydrated and had to be temporarily withdrawn from duty.

Other animals

As well as detecting humans, there are two principal occasions where there is a need to detect animals: first, when surveying wildlife for conservation purposes; second, when looking for infestation or contamination. A wildlife survey may involve searching over large distances in difficult terrain to find a species that is secretive, perhaps nocturnal, and spends a great deal of time underground. The dog must be trained to find the target species, to signal when a find is made and, critically, not to harm the find. The detection may rely on the animal's scent or its faeces or *scat* (Orkin, Yang, Yang, Douglas, & Jiang, 2016).

Cablk and Heaton (2006) describe how detection dogs were used in a survey in the Mojave Desert in California, the driest desert in North America, that focused on the endangered desert tortoise. The analysis of the dogs' performance showed that the dogs were highly effective at detecting the desert tortoises, some as small as 30 mm, both on the surface and in burrows. The detection rate was not influenced by the temperature, which ranged from 12.16° and 26.73°C, the relative humidity which was between 16 percent and 87 percent, or wind speeds up to 8 m/s. There are other successful uses of detection dogs in surveys of threatened animals including brown treesnakes on Guam (Savidge, Stanford, Reed, Haddock, & Yackel Adams, 2011); box turtles in North Carolina, USA (Kapfer, Munñoz, & Tomasek, 2012); Hermann's tortoise in the Mediterranean (Ballouard et al., 2019); koalas in Queensland, Australia (Cristescu et al., 2015); and the pygmy bluetongue lizard in South Australia (Nielsen, Jackson, & Bull, 2016). Some detection dogs work with such finesse that they can discriminate between native and invasive members of the same species (Rosell, Cross, Johnsen, Sundell, & Zedrosser, 2019).

Bennett, Hauser, and Moore (2020) reviewed 61 studies of the use of dogs in conservation work. They suggest that three metrics are needed to evaluate search performance, precision, sensitivity, and effort. As these three are not consistently reported, comparisons across studies are tenuous at best. In addition, human influences such as the experience of the handler should be taken into account. Jamieson, Baxter, and Murray (2018a) make the case that the skills of the handler are critically important but underplayed in the performance of detection dogs. They conducted a questionnaire study, including a personality assessment, of 35 dog handlers in Australia and New Zealand. In terms of personality, the handlers scored high for Agreeableness and low in Neuroticism. The handler's personality was not as important in their success as their training or the bond they formed with their dog. The importance of the dog–handler relationship is emphasised by Jamieson, Baxter, and Murray (2018b), who found that changing a detection in the dog's handler significantly reduced its performance. As with companion animals, the depth of attachment between human and canine is an important consideration in understanding the dynamics of the relationship.

Another use of detection dogs lies in finding animals we see as problematic. The small oval, reddish insect *Cimex lectularius*, better known as the *bed bug*, has plagued humans for millennia (Panagiotakopulu & Buckland, 1999). An adult nocturnal bed bug is typically between 5 and 7 mm in size and does not fly, but it can move rapidly within a room. The female can lay hundreds of eggs, each about the size of a speck of dust, over a lifetime. The bed bug lives on the blood of animals or humans and their bites can leave small red bumps surrounded by blisters in a tell-tale line or zigzag pattern. Needless to say, bed bugs are unwanted and if they infest a hotel, they are a major problem. However, their size and secretiveness makes them difficult to detect, making eradication extremely difficult (Figure 6.4).

Pfiester, Koehler, and Pereira (2008) note that because detector dogs rely on smell rather than vision, they are at an advantage in detecting bed bugs. They reported that dogs trained to locate live bed bugs and bed bug eggs had a 97 percent accuracy rate. However, Cooper, Wang, and Singh (2014) qualified this high detection rate in looking at the performance of different teams of handlers and dogs. They reported a lower mean detection range of 44 percent, but ranging between 10 and 100 percent; this

Figure 6.4 Bed bug, *Cimex lectularius*.
Source: Photograph from Open Pexels.

variation was due to both differences in accuracy between teams and within teams when measured over time.

Bed bugs are not the only small creatures that cause problems for humans and which we need to find for a variety of reasons. In some countries termites, including drywood, powderpost, and dampwood termites, cause costly damage to buildings. Brooks, Oi, and Koehler (2003) reported that trained detector dogs were highly reliable at detecting termite infestations. In various parts of the globe, sniffer dogs also contribute to the detection of threats to agriculture such as the red fire ant (Lin et al., 2011), the red palm weevil (Suma, La Pergola, Longo, & Soroker, 2014), and the redbay ambrosia beetle (Mendel et al., 2018). Sniffer dogs are also used to detect potentially invasive insects within cargo at border crossings (Moser, Brown, Bizo, Andrew, & Taylor, in press).

A study carried out in the Italian Alps was reported by Alasaad et al. (2012), who used dogs trained to detect the scent of Sarcoptes-infected animals to find sick animals and infected carcasses. Sarcoptic mange is a highly contagious skin disease caused by a mite burrowing through the skin, resulting in intense itching and irritation; the scratching causes most of the animal's hair to fall out. The rapid detection and removal of dead animals and treatment of the sick animals, even when they are under snow, is crucial in containing this infectious disease, but is problematic in the wild. The detector dogs successfully assisted in finding the carcasses of mangy wild animals, mainly Alpine ibex and several species of deer, and in the capture of mange-infected wild animals.

Inorganic

There are several types of inorganic substances that humans may have a vested interest in finding, generally because they are illegal, dangerous, or hazardous. These substances are wide and varied; detection dogs are used to find concealed paper currency (Mesloh, Wolf, & Henych, 2002a), to detect human waste contamination in drainage systems (Van De Werfhorst, Murray, Reynolds, Reynolds, & Holden, 2014), and corrosion under the insulation of pipes at gas processing plants (Schoon, Fenjellanger, Kjeldsen, & Goss, 2014). There are, however, two types of inorganic substances of particular concern: explosive material and narcotics.

There are, as recent history shows, many settings where a concealed bomb can cause untold damage and loss of life. The detection of an explosive device before it detonates is an obvious priority. The term *explosive remnant of war (ERW)* refers to unexploded ordnance of which, alongside unexploded shells and cluster bombs, land mines are perhaps the most pernicious. A UN estimate suggests that around the globe, there approximately 100 million unexploded landmines, remnants from wars and other conflicts, in about 80 countries. The counties with the most unexploded land mines, according to WorldAtlas.com, are Egypt (23 million), Iran (16 m), Afghanistan (10 m), Angola (10 m), China (10 m), and Iraq (10 m). Unexploded land mines mean a loss of land for building, farming and other purposes and roads become unused; any animal or human who treads on a mine will be disfigured, lose limbs, or be killed. Children are especially susceptible to ERWs, picking them up to play with and suffering the consequent blindness, deafness, loss of hands, and facial injuries. It is impossible to know the exact figures, but an estimate by the International Campaign to Ban Landmines gives an annual figure in excess of 4,200 people, of which 42 percent are children, who fall victim to ERWs including landmines.

There is a great deal to be gained by clearing landmines, which can remain active for up to 50 years, but there are two obstacles: (1) minefields are intended to be difficult to locate and many have not been identified, while records of where they are placed are likely to be incomplete or non-existent, and new minefields continue to be laid; (2) the cost of surveying and clearing ERWs is highly expensive, particularly as events such as earthquakes, floods, and sandstorms can move ERWs and obliterate or cover any markers. While there is a scientific foundation for the use of dogs as detectors for explosives (Furton & Myers, 2001), in practice, a range of animals – elephants, dogs, rats, honeybees, and wasps – have been used to detect the odours characteristic of volatile substances, including landmines, to aid clearance (Jones, 2011; Lazarowski & Dorman, 2014; Leitch, Anderson, Kirkbride, & Lennard, 2013; Miller et al., 2015; Verhagen, Cox, Machangu, Weetjens, & Billet, 2003).

The detection of illicit narcotics is a continual struggle for law enforcement. The narcotics detection dog is a valuable asset in the work of those agencies, such as police and customs, charged with preventing drug-related offences. The training regime for a narcotics detection dog relies on learning, generally through reward, to detect basic odours – say, cocaine, heroin, marijuana, methamphetamine, and MDMA (ecstasy) – for different types and compositions of narcotics. Once a scent is detected, the dog may be trained to sit or lie down to signal scent recognition. The training of dogs for this type of work necessarily encompasses detection of narcotics contained in luggage and packages or held by people in settings ranging from buildings, cargo areas, vehicles, open areas, and large-scale events such as festivals. In practice, the efficiency of the narcotic search dog depends upon a multitude of factors including the dog's ability, the effectiveness of the training, the skills of the handler, and the nature of the search environment (Hayes, McGreevy, Forbes, Laing, & Stuetz, 2018; Jezierski et al., 2014).

The use of drug detection dogs raises legal issues concerned with the odour detection and police search, intrusion and our rights to privacy, and the admissibility of evidence gained from dogs. Of course, legislation varies from country to country, as illustrated by contributions from America (Bird, 1996, Mesloh, Wolf, & Henych, 2002b), Australia (Lancaster, Ritter, Hughes, & Hoppe, 2017), and the UK (Marks, 2007). When search evidence is permitted in court, the additional problem that arises is of jurors placing undue faith in evidence produced by search dogs (Lit, Oberbauer, Sutton, & Dror, 2019). This issue is akin to the "CSI effect" when jurors, familiar with the scientific wizardry of the crime scene investigators in the television series *CSI*, believe in the near infallibility of forensic science (Podlas, 2006).

Animals at war

It seems that there is nothing more than we humans like than a good fight. In the book *Constant Battles: Why We Fight* (LeBlanc & Register, 2004), the archaeologist Steven LeBlanc writes "The welfare and ecological destruction we find today fit into patterns of human behavior that have gone on for millions of years" (p. xi). As history unfolded, humans became ever more adept at using what was around them, including animals, to improve their chances of winning a fight. Thus, warfare animals have been used as weapons, combatants, transport, and as a means of communication (Salter, 2018). In addition, some sections of the armed forces have an animal as a mascot, which may be an emblem of their unit, or used as part of ceremonies, or for companionship. Trousselard et al. (2014) note that "Dogs are the most common companions for becoming mascots but cats, donkeys, monkeys, lizards, pigs and birds are also adopted as companions and/or mascots" (p. 1822). While a mascot does not take part in conflict, animals can play an active role in warfare. The use of animals as weapons of war was perfected by, amongst others, the ancient Greeks and Romans. In sieges, the attackers catapulted beehives over the city walls to bring swarms of discomfort upon the enemy. It is recorded that in 198 BCE, the Atrenians defended their city of Hatra (in modern-day Iraq) by filling clay pots with scorpions that they flung down on the attacking Romans. In 1591, at the Battle of Tondibi, West Africa, the army of the Songhai Empire used the tactic of charging a thousand stampeding cattle at the Moroccan infantry. This tactic had been successful in past encounters, but this time there was a crucial difference. The Moroccans had guns, the sound of which alarmed the cattle and they turned and stampeded into the Songhai army who lost the battle and ultimately their empire. A not entirely dissimilar tactic was used by the German forces in World War II when they ordered the flooding of the Pontine Marshes south of Rome. These marshes had been drained in the 1920s and 1930s and their reflooding would create a mosquito-infested swamp that would bring malaria and act to slow the Allied advance. There is a debate as to whether this tactic was a form of biological warfare (Geissler & Guillemin, 2010).

In the course of history, some animals became famous for wartime exploits: during the Second Punic War (264 BCE–241 BCE) Hannibal famously led his army of African war elephants over the Alps in his efforts to defeat the Romans. While many elephants died in the harsh conditions, those that survived took part Hannibal's victory in the Battle of Trebia where they caused panic among the Roman cavalry. In Battle of Balaclava during the Crimean War (1853–1856), the British light cavalry charged against heavily defended Russian forces, a calamitous mistake brought about by the Battle misinformation and miscommunication, and were valiantly slaughtered. In his 1854 poem, "The Charge of the Light Brigade," Alfred, Lord Tennyson observed that:

> Cannon behind them
> Volleyed and thundered;
> Stormed at with shot and shell,
> While horse and hero fell.

In 1942 a group of Polish II Corps soldiers, evacuated from the Soviet Union during World War II, were at a railway station in Hamadan, Iran, where they purchased a bear cub, a Syrian brown bear. They called the cub Wojtek (the name means "one who enjoys as a soldier") and he became part of the soldiers' group. He learned to behave like a soldier supplementing his diet of fruit, honey and marmalade with coffee, cigarettes (which he ate), and beer, his favourite tipple. Wojtek imitated the other soldiers learning to salute when greeted, marching on his hind legs and carrying ammunition crates. He grew up on campaign and went with Polish II Corps to fight in the Italian campaign with the British Eighth Army. The British ships taking the troops from Egypt to Italy did not carry animals so to circumvent this restriction, Wojtek was enlisted as a private in the Polish Army, with his own paybook and serial number, and numbered among the soldiers of the 22nd Artillery Supply Company.

Wojtek saw action including the Battle of Monte Cassino in 1944 where, the (apocryphal) story is that Wojtek worked in the ammunition supply, carrying 100-pound crates of shells. It is true that he copied the soldiers lifting crates and his strength allowed him to carry loads that normally required several men so he may have done this in combat. His efforts earned him promotion to the rank of corporal. Wojtek's value was reflected in the adoption of a bear carrying an artillery shell as the official emblem of the

22nd Company. In 1945, at the end of World War II, the 22nd Company was transported to Berwickshire where Wojtek became popular and an honorary member of the Polish-Scottish Association. Following demobilisation in November 1947, Wojtek was given to Edinburgh Zoo – his visitors included former Polish soldiers who would throw him cigarettes to eat – where he lived until his death 1963 at the age of 21 (see Forsyth, 2017). The media attention made Wojtek and popular figure and he appeared on television in the *Blue Peter* programme, and in 2011 a film about him, *Wojtek: The Bear That Went to War*, was released. There are statues of Wojtek in the town of Duns, Berwickshire, in Krakow, Poland, and in Princes Street Gardens, Edinburgh (Figure 6.5).

While Wojtek is something of a cause célèbre, he is not alone in becoming a national hero. For example, the German Shepherd Max is widely known throughout Romania as a war hero who saw action with the army in Afghanistan. He died in 2015 and was buried with military honours in the Animals' Heaven Cemetery near Bucharest, as now can be seen on Facebook and YouTube (Rujoiu & Rujoiu, 2018).

The pigeon became an unlikely instrument of conflict in World War II with the advent of *Project Pigeon*, B. F. Skinner's research programme designed to use operant learning to train pigeons to guide a bomb [see Schultz-Figueroa (2019) for an account of this project and the wider issues it raises regarding our use of animals]. A pigeon trained to recognise a target was placed in front of a screen and, on seeing the target, it pecked at

Figure 6.5 Statute of Wojtek, Princess Street Gardens, Edinburgh.
Source: Photograph by Taras Young. Creative Commons Attribution-Share Alike 4.0 International License.

the screen. When the target was in the centre of the screen, the screen remained stationary, but if the bomb started to go off track the image moved towards the edge of the screen. The pigeon, trained to track the image, pecked at the moving image, in turn moving the screen on its pivots. Thus, movement was detected by sensors which then signalled the control mechanism and brought the bomb back on track. This process continued until the bomb reached its target and detonated, pigeon and all. The project never reached fruition and was cancelled by the Army in 1944.

Since the 1960s the US Navy has used bottlenose dolphins in military work, training the dolphins to recognise objects underwater that are imperceptible to human divers. The Navy dolphins work with human handlers patrolling Navy harbours and other shipping areas where the dolphin's natural echolocation allows it to detect threats such as marine mines and limpet bombs that attach to the hulls of warships. If a dolphin detects a strange object, it signals "yes" and the handler can act accordingly. The dolphins are also trained to detect enemy divers and to help people in difficulty in the water. Alongside dolphins, California sea lions are put to use by the US Navy: the prize assets of the sea lion are exceptional low-light vision and underwater hearing, fast swimming speeds of up to 25 mph, and diving to depths of up to 1,000 feet. The sea lions locate marine mines and enemy divers and swimmers and can be equipped with cameras to provide live underwater pictures.

In war, a great deal of equipment – firearms, ammunition, food and water, fuel, and so on – must be carried from place to place as the armies move. While the modern army has mechanised transport, in centuries past the beasts of burden were mules, donkeys, and heavy horses; in deserts, the camel assumed this role. The mule, a hybrid of a male donkey and a female horse, was prized for its mild nature, endurance, sure-footedness, and ability to carry heavy loads. The Romans used a single mule to carry the belongings and supplies of eight soldiers. Napoleon Bonaparte rode on a mule as he led the French armies across the Alps into Italy. In World War I, the American Army used over one-half million horses and mules, with almost 70,000 killed in action. The use of mules in difficult terrain continues to this day: the American forces in Afghanistan used mules to help keep open supply lines to remote mountain bases.

Finally, good communication is a critical aspect of warfare, quick and efficient communication enables timely battlefield responses; poor communication leads to expensive disasters. The animal that proved to be highly adept at carrying messages in time of war was the homing or carrier pigeon. The first recorded use of the pigeon for message delivery – the written messages were typically carried in a small cylinder tied to the pigeon's leg – was in Egypt in 2900 BCE, then subsequently by the Greeks and the Romans and in the Crusades. In World War I, pigeons were used by many warring countries, including Belgium, France, England, Germany, and the USA, employed mainly by artillery, cavalry, infantry, and tank units but also by the navy and air force (Katzung Hokanson, 2018; Phillips, 2018). In 1938, the UK National Pigeon Service supplied hundreds of thousands of pigeons for use in World War II. Innumerable pigeons were killed by artillery fire and by gas and some were decorated for their exploits under fire.

Acts of valour

In war, some people commit exceptional and extreme acts of valour, showing courage in the face of enemy fire or taking apparently insurmountable risks to save lives. These acts may be recognised by the award of a medal, with different medals for different levels of courage. Thus, for example, in Britain the *Victoria Cross* is the highest award for valour "in the presence of the enemy"; France have the *Croix de Guerre*; and the USA awards the Medal of Honour (also called the Congressional Medal of Honour). It is unclear exactly why some people carry out acts of heroism: maybe their actions are instinctive or impulsive, so allowing little or no decision making, or because they have no wish to see another person die alone (see Kugel, Hausman, Black, & Bongar, 2017). Can an animal act with courage in the presence of the enemy? Whether animals are capable of courageous acts has been debated since the ancient Greeks, but that has not prevented animals being awarded for conspicuous valour.

In World War I, the American 77th Division came under "friendly fire" from its own artillery. The commanding officer sent pigeons to base but saw them fall under fire: the last pigeon, called *Cher Ami*, was sent in a final bid to stop the barrage. The bird flew towards German fire and was hit and fell to the ground only to struggle back into the air: 30 minutes and 25 miles later he arrived at base, heavily wounded, and the message he carried resulted in the artillery taking up new firing coordinates, saving the lives of many soldiers. The medics saved *Cher Ami* and for his gallantry in the field he was awarded the *Croix de Guerre with Palm*, one of France's highest military honours. *Cher Ami* died in 1919 in New Jersey, where his body was preserved for presentation to the American government.

Animals decorated in World War I received medals that could also be awarded to human military. However, during World War II, Maria Dickin, who also founded the People's Dispensary for Sick Animals (PDSA), instituted a medal – a bronze medallion, inscribed "For Gallantry" and "We Also Serve" – to honour animals in war. The Dickin Medal was awarded 54 times between 1943 and 1949: the recipients were 32 pigeons, 18 dogs, 3 horses, and a cat.

Gone but not forgotten

In warfare, people and animals die. In modern times, there have been attempts to formulate standards of international law for the humanitarian treatment of those involved in war. The Geneva Conventions of 1949 details the basic rights of: (1) the "Wounded and Sick in Armed Forces in the Field"; (2) the "Wounded, Sick and Shipwrecked Members of Armed Forces at Sea"; (3) the "Treatment of Prisoners of War"; and (4) the "Protection of Civilian Persons in Time of War" (see van Dijk, 2018). de Hemptinne (2020) makes the point that: "Being deeply anthropocentric, international humanitarian law (IHL) largely ignores the protection of animals" (p. 174). As animals serve their country, suffer injury and death, and can even be taken as prisoners of war – see Nowrot (2015) for the story of the British military working dog taken prisoner by the Taliban in Afghanistan – so de Hemptinne argues that given modern-day understanding of animal sentience, it is incumbent upon humans to make the necessary changes to the law.

Honouring the dead

Many cities, towns, and villages have war memorials to honour those who fell during conflict, but the portrayal of animals is invisible or at best secondary to humans (Johnston, 2012; Kean, 2013). This neglect of the role of animals in warfare has recently shifted with the erection of memorials to animals killed in war; for example, in 1994 the

Marine War Dog Memorial was unveiled at the United States Marine Corps War Dog Cemetery dedicated to the 25 dogs killed "liberating Guam in 1944"; in 2004 in Hyde Park, London, there is the Animals in War Memorial. These memorials have attracted vandalism and graffiti that have a political motivation or some other purpose. In commenting on the vandalism to the memorial in Hyde Park, Wilson (2018) observes that: "More notable still, the vandals drew red lines through the central inscription, which explains the logic behind its commemoration. The spray paint thus made the following language nearly illegible: 'This monument is dedicated to all the animals that served and died alongside British and Allied Forces in wars and campaigns throughout time'" (p. 90).

References

Alasaad, S., Permunian, R., Gakuya, F., Mutinda, M., Soriguer, R. C., & Rossi, L. (2012). Sarcoptic-mange detector dogs used to identify infected animals during outbreaks in wildlife. *BMC Veterinary Research*, *8*, 110.

Aldrich, B. C. (2018). The use of primate "actors" in feature films 1990–2013. *Anthrozoös*, *31*, 5–21.

Audrestch, H. M., Whelan, C. T., Grice, D., Asher, L., England, G. C., & Freeman, S. L. (2015). Recognizing the value of assistance dogs in society. *Disability and Health Journal*, *8*, 469–474.

Ballouard, J.-M., Raphael, G., Fabien, R., Aurelien, B., Sebastien, C., Bech, N., & Xavier, B. (2019). Excellent performances of dogs to detect cryptic tortoises in Mediterranean scrublands. *Biodiversity and Conservation*, *28*, 4027–4045.

Bashaw, M. J., Gibson, M. D., Schowe, D. M., & Kucher, A. S. (2016). Does enrichment improve reptile welfare? Leopard geckos (*Eublepharis macularius*) respond to five types of environmental enrichment. *Applied Animal Behaviour Science*, *184*, 150–160.

Baston, K. (2018). Circus at the edge of Europe: Acrobatic entertainments in the Ottoman Empire. *Early Popular Visual Culture*, *16*, 57–82.

Batt, L., Batt, M., Baguley, J., & McGreevy, P. (2010). Relationships between puppy management practices and reported measures of success in guide dog training. *Journal of Veterinary Behavior*, *5*, 240–246.

Batt, L. S., Batt, M. S., Baguley, J. A., & McGreevy, P. D. (2008). Factors associated with success in guide dog training. *Journal of Veterinary Behavior*, *3*, 143–151.

Baxter, C. L., & Hargrave, M. L. (2015). *Guidance on the use of historic human remains detection dogs for locating unmarked cemeteries*. Champaign, IL: US Army Engineer Research and Development Center.

Bennett, E. M., Hauser, C. E., & Moore, J. L. (2020). Evaluating conservation dogs in the search for rare species. *Conservation Biology*, *34*, 314–325.

Bird, R. C. (1996). An examination of the training and reliability of the narcotics detection dog. *Kentucky Law Journal*, *85*, 405–433.

Blau, S., Sterenberg, J., Weeden, P., Urzedo, F., Wright, R., & Watson, C. (2018). Exploring non-invasive approaches to assist in the detection of clandestine human burials: Developing a way forward. *Forensic Sciences Research*, *3*, 320–342.

Bogaerts, E., Moons, C. P. H., Van Nieuwerburgh, F., Peelman, L., Saunders, J. H., & Broeckx, B. J. G. (2019). Rejections in a non-purpose bred assistance dog population: Reasons, consequences and methods for screening. *PLoS One*, *14*, e0218339.

Brando, S., & Buchanan-Smith, H. M. (2018). The 24/7 approach to promoting optimal welfare for captive wild animals. *Behavioural Processes*, *156*, 83–95.

Brelsford, V., Meints, K., Gee, N., & Pfeffer, K. (2017). Animal-assisted interventions in the classroom — A systematic review. *International Journal of Environmental Research and Public Health*, *14*, 669.

Brooks, S. E., Oi, F. M., & Koehler, P. G. (2003). Ability of canine termite detectors to locate live termites and discriminate them from non-termite material. *Journal of Economic Entomology*, *96*, 1259–1266.

Brown, T. J., Ham, S. H., & Hughes, M. (2010). Picking up litter: An application of theory-based communication to influence tourist behaviour in protected areas. *Journal of Sustainable Tourism, 18,* 879–900.

Cablk, M. E., & Heaton, J. S. (2006). Accuracy and reliability of dogs in surveying for desert tortoise (*Gopherus Agassizii*). *Ecological Applications, 16,* 1926–1935.

Cablk, M. E., & Sagebiel, J. C. (2011). Field capability of dogs to locate individual human teeth. *Journal of Forensic Sciences, 56,* 1018–1024.

Choo, Y., Todd, P. A., & Li, D. (2011). Visitor effects on zoo orangutans in two novel, naturalistic enclosures. *Applied Animal Behaviour Science, 133,* 78–86.

Cless, I. T., & Lucas, K. E. (2017). Variables affecting the manifestation of and intensity of pacing behavior: A preliminary case study in zoo-housed polar bears. *Zoo Biology, 36,* 307–315.

Cole, K., & Fraser, D. (2018). Zoo animal welfare: The human dimension. *Journal of Applied Animal Welfare Science, 21,* 49–58.

Collins, C., Quirke, T., McKeown, S., Flannery, K., Kennedy, D., & O'Riordan, R. (2019). Zoological education: Can it change behaviour? *Applied Animal Behaviour Science, 220,* Article 104857.

Cook, S., & Hosey, G. R. (1995). Interaction sequences between chimpanzees and human visitors at the zoo. *Zoo Biology, 14,* 431–440.

Cooper, R., Wang, C., & Singh, N. (2014). Accuracy of trained canines for detecting bed bugs (Hemiptera: Cimicidae). *Journal of Economic Entomology, 107,* 2171–2181.

Craigon, P. J., Hobson-West, P., England, G. C. W., Whelan, C., Lethbridge, E., & Asher, L. (2017). "She's a dog at the end of the day": Guide dog owners' perspectives on the behaviour of their guide dog. *PLoS One, 12,* e0176018.

Cristescu, R. H., Foley, E., Markula, A., Jackson, G., Jones, D., & Frère, C. (2015). Accuracy and efficiency of detection dogs: A powerful new tool for koala conservation and management. *Scientific Reports, 5,* Article Number 8349.

Davey, G. (2007). Visitors' effects on the welfare of animals in the zoo: A review. *Journal of Applied Animal Welfare Science, 10,* 169–183.

de Hemptinne, J. (2020). Challenges regarding the protection of animals during warfare. In A. Peters (Ed.), *Studies in Global Animal Law* (pp. 173–183). Berlin: Springer Open.

Dellinger, M. (2009). Using dogs for emotional support of testifying victims of crime. *Animal Law Review, 15,* 171–192.

DeMatteo, K. E., Davenport, B., & Wilson, L. E. (2019). Back to the basics with conservation detection dogs: Fundamentals for success. *Wildlife Biology, 1,* 1–9.

de Vere, A. J. (2018). Visitor effects on a zoo population of California sea lions (*Zalophus californianus*) and harbor seals (*Phoca vitulina*). *Zoo Biology, 37,* 162–170.

Don, M. (2016). *Nigel: My family and other dogs.* London: Hatchette.

Draper, C., Browne, W., & Harris, S. (2013). Do formal inspections ensure that British zoos meet and improve on minimum animal welfare standards? *Animals, 3,* 1058–1072.

Esson, M., & Moss, A. G. (2014). Zoos as a context for reinforcing environmentally responsible behaviour: The dual challenges that zoo educators have set themselves. *Journal of Zoo and Aquarium Research, 2,* 8–13.

Fairchild, T. P., Fowler, M. S., Pahl, S., & Griffin, J. N. (2018). Multiple dimensions of biodiversity drive human interest in tide pool communities. *Scientific Reports, 8,* Article Number 15234.

Falk, J. H., & Adelman, L. M. (2003). Investigating the impact of prior knowledge and interest on aquarium visitor learning. *Journal of Research in Science Teaching, 40,* 163–176.

Fife-Cook, I., & Franks, B. (2019). Positive welfare for fishes: Rationale and areas for future study. *Fishes, 4,* 31.

Fishman, G. A. (2003). When your eyes have a wet nose: The evolution of the use of guide dogs and establishing the seeing eye. *Survey of Ophthalmology, 48,* 452–458.

Forsyth, I. (2017). A bear's biography: Hybrid warfare and the more-than-human battlespace. *Environment and Planning D: Society and Space, 35,* 495–512.

Fraser, J. (2009). The anticipated utility of zoos for developing moral concern in children. *Curator, 52,* 349–361.

Frynta, D., Peléšková, Š., Rádlová, S., Janovcová, M., & Landová, E. (2019). Human evaluation of amphibian species: A comparison of disgust and beauty. *The Science of Nature, 106*(7–8), 41.

Frynta, D., Šimková, O., Lišková, S., & Landová, E. (2013). Mammalian collection on Noah's Ark: The effects of beauty, brain and body size. *PLoS One, 8,* e63110.

Fung, S. C. (2017). Canine-assisted reading programs for children with special educational needs: Rationale and recommendations for the use of dogs in assisting learning. *Educational Review, 69,* 435–450.

Furton, K. G., & Myers, L. J. (2001). The scientific foundation and efficacy of the use of canines as chemical detectors for explosives. *Talanta, 54,* 487–500.

Geissler, E., & Guillemin, J. (2010). German flooding of the Pontine Marshes in World War II: Biological warfare or total war tactic? *Politics and the Life Sciences, 29,* 2–23.

Gill, S. R., & Warrington, W. (2017). Sustainability initiatives in zoos and aquariums: Looking in to reach out. *Leisure/Loisir, 41,* 443–465.

Glavaš, V., & Pintar, A. (2019). Human remains detection dogs as a new prospecting method in archaeology. *Journal of Archaeological Method and Theory, 26,* 1106–1124.

Goodenough, A. E., McDonald, K., Moody, K., & Wheeler, C. (2019). Are "visitor effects" overestimated? Behaviour in captive lemurs is mainly driven by co-variation with time and weather. *Journal of Zoo and Aquarium Research, 7,* 59–66.

Gravrok, J., Bendrups, D., Howell, T., & Bennett, P. (2019). Beyond the benefits of assistance dogs: Exploring challenges experienced by first-time handlers. *Animals : An Open Access Journal from MDPI, 9,* 203.

Greco, B. J., Meehan, C. L., Miller, L. J., Shepherdson, D. J., Morfeld. K. A., & Andrews, J., ... Mench, J. A. (2016). Elephant management in North American zoos: Environmental enrichment, feeding, exercise, and training. *PLoS One, 11,* e0152490.

Hall, S. S., Gee, N. R., & Mills, D. S. (2016). Children reading to dogs: A systematic review of the literature. *PLoS One, 11,* e0149759.

Harvey, N. D., Craigon, P. J., Sommerville, R., McMillan, C., Green, M., England, G. C. W., & Asher, L. (2016). Test-retest reliability and predictive validity of a juvenile guide dog behavior test. *Journal of Veterinary Behavior, 11,* 65–76.

Hayes, J. E., McGreevy, P. D., Forbes, S. L., Laing, G., & Stuetz, R. M. (2018). Critical review of dog detection and the influences of physiology, training, and analytical methodologies. *Talanta, 185,* 499–512.

Hitchens, P. L., Hultgren, J., Frössling, J., Emanuelson, U., & Keeling, L. J. (2017). Circus and zoo animal welfare in Sweden: An epidemiological analysis of data from regulatory inspections by the official competent authorities. *Animal Welfare, 26,* 373–382.

Hosey, G. R. (2000). Zoo animals and their human audiences: What is the visitor effect? *Animal Welfare, 9,* 343–357.

Hosey, G., & Melfi, V. (2012). Human–animal bonds between zoo professionals and the animals in their care. *Zoo Biology, 31,* 13–26.

Hosey, G., & Melfi, V. (2015). Are we ignoring neutral and negative human–animal relationships in zoos? *Zoo Biology, 34,* 1–8.

Iossa, G., Soulsbury, C. D., & Harris, S. (2009). Are wild animals suited to a travelling circus life? *Animal Welfare, 18,* 129–140.

Jalongo, M. R. (2005). "What are all these Dogs Doing at School?": Using therapy dogs to promote children's reading practice. *Childhood Education, 81,* 152–158.

Jamieson, L. T. J., Baxter, G. S., & Murray, P. J. (2018a). Who's a good handler? Important skills and personality profiles of wildlife detection dog handlers. *Animals : An Open Access Journal from MDPI, 8,* 222.

Jamieson, L. T. J., Baxter, G. S., & Murray, P. J. (2018b). You Are Not My Handler! Impact of changing handlers on dogs' behaviours and detection performance. *Animals: An Open Access Journal from MDPI, 8,* 176.

Janovcová, M., Rádlová, S., Polák, J., Sedláčková, K., Pelešková, S., Žampachová, B., Frynta, D., & Landová, E. (2019). Human attitude toward reptiles: A relationship between fear, disgust, and aesthetic preferences. *Animals : An Open Access Journal from MDPI, 9*, 238.

Jezierski, T., Adamkiewicz, E., Walczak, M., Sobczyńska, M., Górecka-Bruzda, A., Ensminger, J., & Papet, E. (2014). Efficacy of drug detection by fully-trained police dogs varies by breed, training level, type of drug and search environment. *Forensic Science International, 237*, 112–118.

Johnston, S. (2012). Animals in war: Commemoration, patriotism, death. *Political Research Quarterly, 65*, 359–371.

Jones, B. M. (2011). Applied behavior analysis is ideal for the development of a land mine detection technology using animals. *The Behavior Analyst, 34*, 55–73.

Kastelein, R. A., Jennings, N., & Postma, J. (2007). Feeding enrichment methods for pacific walrus calves. *Zoo Biology, 26*, 175–186.

Katzung Hokanson, B. R. (2018). Saving grace on feathered wings: Homing pigeons in the First World War. *The Gettysburg Historical Journal, 17*, Article 7.

Kean, H. (2013). Animals and war memorials: Different approaches to commemorating the human-animal relationship. In R. Hediger (Ed.), *Animals and war: Studies of Europe and North America* (pp. 237–262). Leiden, The Netherlands: Brill Publishing.

Komar, D. (1999). The use of cadaver dogs in locating scattered, scavenged human remains: Preliminary field test results. *Journal of Forensic Science, 44*, 405–408.

Kopczak, C., Kisiel, J. F., & Rowe, S. (2015). Families talking about ecology at touch tanks. *Environmental Education Research, 21*, 129–144.

Krebs, B. L., Marrin, D., Phelps, A., Krol, L., & Watters, J. V. (2018). Managing aged animals in zoos to promote positive welfare: A review and future directions. *Animals: An Open Access Journal from MDPI, 8*, 116.

Kugel, U., Hausman, C., Black, L., & Bongar, B. (2017). *Psychology of physical bravery*. Oxford: Oxford Handbooks Online.

La Toya, J. J., Baxter, G. S., & Murray, P. J. (2017). Identifying suitable detection dogs. *Applied Animal Behaviour Science, 195*, 1–7.

Lancaster, K., Ritter, A., Hughes, C., & Hoppe, R. (2017). A critical examination of the introduction of drug detection dogs for policing of illicit drugs in New South Wales, Australia using Kingdon's 'multiple streams' heuristic. *Evidence & Policy: A Journal of Research, Debate and Practice, 13*, 583–603.

Larson, D. O., Vass, A. A., & Wise, M. (2011). Advanced scientific methods and procedures in the forensic investigation of clandestine graves. *Journal of Contemporary Criminal Justice, 27*, 149–182.

Lasseter, A. E., Jacobi, K. P., Farley, R., & Hensel, L. (2003). Cadaver dog and handler team capabilities in the recovery of buried human remains in the Southeastern United States. *Journal of Forensic Science, 48*, 1–5.

Lavers, K. (2015). Horses in modern, new, and contemporary circus. *Animal Studies Journal, 4*, 140–172.

Lazarowski, L., & Dorman, D. C. (2014). Explosives detection by military working dogs: Olfactory generalization from components to mixtures. *Applied Animal Behaviour Science, 151*, 84–93.

Lazarowski, L., Haney, P. S., Brock, J., Fischer, T., Rogers, B., Angle, C., ... Waggoner, L. P. (2018). Investigation of the behavioral characteristics of dogs purpose-bred and prepared to perform *Vapor Wake*® detection of person-borne explosives. *Frontiers in Veterinary Science, 5*, Article 50.

LeBlanc, S. A., & Register, K. E. (2004). *Constant battles: Why we fight*. New York: St. Martin's Griffin.

Leitch, O., Anderson, A., Kirkbride, K. P., & Lennard, C. (2013). Biological organisms as volatile compound detectors: A review. *Forensic Science International, 232*, 92–103.

Li, K., Kou, J., Lam, Y., Lyons, P., & Nguyen, S. (2019). First-time experience in owning a dog guide by older adults with vision loss. *Journal of Visual Impairment & Blindness, 113*, 452–463.

Lin, H. M., Chi, W. L., Lin, C. C., Tseng, Y. C., Chen, W. T., Kung, Y. L., ... & Chen, Y. Y. (2011). Fire ant-detecting canines: A complementary method in detecting red imported fire ants. *Journal of Economic Entomology, 104*, 225–231.

Lit, L., Oberbauer, A., Sutton, J. E., & Dror, I. E. (2019). Perceived infallibility of detection dog evidence: Implications for juror decision-making. *Criminal Justice Studies, 32*, 189–206.

Lit, L., Schweitzer, J. B., & Oberbauer, A. M. (2011). Handler beliefs affect scent detection dog outcomes. *Animal Cognition*, *14*, 387–394.

Luebke, J. F. (2018). Zoo exhibit experiences and visitors' affective reactions: A preliminary study. *Curator: The Museum Journal*, *61*, 345–352.

Luebke, J. F., Watters, J. V., Packer, J., Miller, L. J., & Powell, D. M. (2016). Zoo visitors' affective responses to observing animal behaviors. *Visitor Studies*, *19*, 60–76.

Maple, T. L., & Perdue, B. M. (2013). Environmental enrichment. In T. L. Maple & B. M. Perdue (Eds.), *Zoo animal welfare* (pp. 95–117). Berlin: Springer.

Marks, A. (2007). Drug detection dogs and the growth of olfactory surveillance: Beyond the rule of law? *Surveillance & Society*, *4*, 257–271.

Martellucci, S., Belvisi, V., Ralli, M., Stadio, A. D., Musacchio, A., Greco, A., ... Attansio, G. (2019). Assistance dogs for persons with hearing impairment: A review. *International Tinnitus Journal*, *23*, 26–30.

Mason, G. (2006). Stereotypic behaviour in captive animals: Fundamentals and implications for welfare and beyond. In G. Mason & J. Rushen (Eds.), *Stereotypic animal behaviour: Fundamentals and applications to welfare* (2nd ed., pp. 325–356). Wallingford, Oxfordshire: CABI.

Mason, G. J. (1991). Stereotypies and suffering. *Behavioural Processes*, *25*, 103–115.

Mason, G. J., & Latham, G. J. (2004). Can't stop won't stop: Is stereotypy a reliable animal welfare indicator? *Animal Welfare*, *13*, S57–S69.

Mellish, S., Pearson, E. L., McLeod, E. M., Tuckey, M. R., & Ryan, J. C. (2019). What goes up must come down: An evaluation of a zoo conservation-education program for balloon litter on visitor understanding, attitudes, and behaviour. *Journal of Sustainable Tourism*, *27*, 1393–1415.

Mendel, J., Burns, C., Kallifatidis, B., Evans, E., Crane, J., Furton, K. G., & Mills, D. (2018). Agri-dogs: Using canines for earlier detection of laurel wilt disease affecting avocado trees in South Florida. *HortTechnology*, *28*, 109–116.

Mesloh, C., Wolf, R., & Henych, M. (2002a). Sniff test; Utilization of the law enforcement canine in the seizure of paper currency. *Journal of Forensic Identification*, *52*, 704–724.

Mesloh, C., Wolf, R., & Henych, M. (2002b). Scent as forensic evidence and its relationship to the law enforcement canine. *Journal of Forensic Identification*, *52*, 169–182.

Migala, A. F., & Brown, S. E. (2012). Use of human remains detection dogs for wide area search after wildfire: A new experience for Texas Task Force 1 search and rescue resources. *Wilderness & Environmental Medicine*, *23*, 337–342.

Miller, A. K., Hensman, M. C., Hensman, S., Schultz, K., Reid, P., Shore, M., ... Stephen Lee. (2015). African elephants (*Loxodonta africana*) can detect TNT using olfaction: Implications for biosensor application. *Applied Animal Behaviour Science*, *171*, 177–183.

Morgan, K. N., & Tromborg, C. T. (2007). Sources of stress in captivity. *Applied Animal Behaviour Science*, *102*, 262–302.

Moser, A. Y., Brown, W. Y., Bizo, L. A., Andrew, N., & Taylor, M. K. (in press). Biosecurity dogs detect live insects after training with odour-proxy training aids: Scent extract and dead specimens. *Chemical Senses*, *45*, 179–186.

Moss, A., & Esson, M. (2010). Visitor interest in zoo animals and the implications for collection planning and zoo education programmes. *Zoo Biology*, *29*, 715–731.

Kapfer, J. M., Munñoz, D. J., & Tomasek, T. (2012). Use of wildlife detector dogs to study Eastern Box Turtle (*Terrapene Carolina Carolina*) Populations. *Herpetological Conservation and Biology*, *7*, 169–175.

Mäekivi, N. (2018). Freedom in captivity: Managing zoo animals according to the 'five freedoms'. *Biosemiotics*, *11*, 7–25.

Myers Jr, O. E., Saunders, C. D., & Birjulin, A. A. (2004). Emotional dimensions of watching zoo animals: An experience sampling study building on insights from psychology. *Curator: The Museum Journal*, *47*, 299–321.

Naylor, W., & Parsons, E. C. M. (2019). An international online survey on public attitudes towards the keeping of whales and dolphins in captivity. *Tourism in Marine Environments*, *14*, 133–142.

Neumann, D. L., Chan, R. C., Boyle, G. J., Wang, Y., & Westbury, H. R. (2015). Measures of empathy: Self-report, behavioral, and neuroscientific approaches. In G. J. Boyle, D. H. Saklofske & G. Matthews (Eds.), *Measures of personality and social psychological constructs* (pp. 257–289). Amsterdam, The Netherlands: Elsevier/Academic Press.

Nielsen, T. P., Jackson, G., & Bull, C. M. (2016). A nose for lizards: Can a detection dog locate the endangered pygmy bluetongue lizard (*Tiliqua adelaidensis*)? *Transactions of the Royal Society of South Australia, 140*, 234–243.

Nowrot, K. (2015). Animals at war: The status of "animal soldiers" under international humanitarian law. *Historical Social Research, 40*, 128–150.

Ogle, B. (2016). Value of guest interaction in touch pools at public aquariums. *Universal Journal of Management, 4*, 59–63.

Orkin, J. D., Yang, Y., Yang, C., Douglas, W. Y., & Jiang, X. (2016). Cost-effective scat-detection dogs: Unleashing a powerful new tool for international mammalian conservation biology. *Scientific Reports, 6*, Article Number 34758.

Panagiotakopulu, E., & Buckland, P. C. (1999). *Cimex lectularius* L., the common bed bug from pharaonic Egypt. *Antiquity, 73*, 908–911.

Parenti, L., Foreman, A., Meade, B. J., & Wirth, O. (2013). A revised taxonomy of assistance animals. *Journal of Rehabilitation Research and Development, 50*, 745–756.

Patoka, J., Vejtrubová, M., Vrabec, V., & Masopustová, R. (2018). Which wild aardvarks are most suitable for outdoor enclosures in zoological gardens in the European Union? *Journal of Applied Animal Welfare Science, 21*, 1–7.

Paul, E. S. (2000). Empathy with animals and with humans: Are they linked? *Anthrozoös, 13*, 194–202.

Pavitt, B., & Moss, A. G. (2019). Assessing the effect of zoo exhibit design on visitor engagement and attitudes towards conservation. *Journal of Zoo and Aquarium Research, 7*, 186–194.

Pedersen, J., Sorensen, K., Lupo, B., & Marx, L. (2019). Human–ape interactions in a zoo setting: Gorillas and orangutans modify their behavior depending upon human familiarity. *Anthrozoös, 32*, 319–332.

Pfattheicher, S., Sassenrath, C., & Schindler, S. (2016). Feelings for the suffering of others and the environment: Compassion fosters proenvironmental tendencies. *Environment and Behavior, 48*, 929–945.

Pfiester, M., Koehler, P. G., & Pereira, R. M. (2008). Ability of bed bug-detecting canines to locate live bed bugs and viable bed bug eggs. *Journal of Economic Entomology, 101*, 1389–1396.

Phillips, C., & Peck, D. (2007). The effects of personality of keepers and tigers (*Panthera tigris tigris*) on their behaviour in an interactive zoo exhibit. *Applied Animal Behaviour Science, 106*, 244–258.

Phillips, G. (2018). Pigeons in the trenches: Animals, communications technologies and the British Expeditionary Force, 1914-1918. *British Journal for Military History, 4*, 60–80.

Podlas, K. (2006). The CSI effect and other forensic fictions. *Loyola of Los Angeles Entertainment Law Review, 27*, 87–125.

Queiroz, M. B., & Young, R. J. (2018). The different physical and behavioural characteristics of zoo mammals that influence their response to visitors. *Animals, 8*, 139.

Quirke, T., & O'Riordan, R. M. (2011). The effect of different types of enrichment on the behaviour of cheetahs (*Acinonyx jubatus*) in captivity. *Applied Animal Behaviour Science, 133*, 87–94.

Rintala, D. H., Matamoros, R., & Seitz, L. L. (2008). Effects of assistance dogs on persons with mobility or hearing impairments: A pilot study. *Journal of Rehabilitation Research & Development, 45*, 489–504.

Rooney, N. J., Bradshaw, J. W. S., & Almey, H. (2004). Attributes of specialist search dogs – A questionnaire survey of UK dog handlers. *Journal of Forensic Science, 2*, 1–7.

Rose, P. E., Brereton, J. E., & Croft, D. P. (2018). Measuring welfare in captive flamingos: Activity patterns and exhibit usage in zoo-housed birds. *Applied Animal Behaviour Science, 205*, 115–125.

Rose, P. E., Nash, S. M., & Riley, L. M. (2017). To pace or not to pace? A review of what abnormal repetitive behavior tells us about zoo animal management. *Journal of Veterinary Behavior, 20*, 11–21.

Rosell, F., Cross, H. B., Johnsen, C. B., Sundell, J., & Zedrosser, A. (2019). Scent-sniffing dogs can discriminate between native Eurasian and invasive North American beavers. *Scientific Reports, 9*, Article Number 15952.

Rujoiu, O., & Rujoiu, V. (2018). Losing Max, the first Romanian Army's military working dog hero. *Journal of Loss and Trauma, 23*, 608–621.

Salter, C. (2018). Animals in the military. In C. G. Scanes & S. R. Toukhsati (Eds.), *Animals and human society* (pp. 195–223). London: Elsevier, Academic Press.

Savidge, J. A., Stanford, J. W., Reed, R. N., Haddock, G. R., & Yackel Adams, A. A. (2011). Canine detection of free-ranging brown treesnakes on Guam. *New Zealand Journal of Ecology, 35*, 174–181.

Scanes, C. G. (2018). Animals in entertainment. In C. G. Scanes & S. R. Toukhsati (Eds.), *Animals and human society* (pp. 225–255). Amsterdam, The Netherlands: Elsevier.

Schoon, A., Fenjellanger, R., Kjeldsen, M., & Goss, K. (2014). Using dogs to detect hidden corrosion. *Applied Animal Behaviour Science, 153*, 43–52.

Schroepfer, K. K., Rosati, A. G., Chartrand, T., & Hare, B. (2011). Use of "entertainment" chimpanzees in commercials distorts public perception regarding their conservation status. *PLoS One, 6*, e26048.

Schultz, J. Y., & Young, J. K. (2018). Behavioral and spatial responses of captive coyotes to human activity. *Applied Animal Behaviour Science, 205*, 83–88.

Schultz-Figueroa, B. (2019). Project Pigeon: Rendering the war animal through optical technology. *JCMS: Journal of Cinema and Media Studies, 58*, 92–111.

Serpell, J. A., & Hsu, Y. (2001). Development and evaluation of a novel method for evaluating behavior and temperament in guide dogs. *Applied Animal Behaviour Science, 72*, 347–364.

Sherwen, S. L., & Hemsworth, P. H. (2019). The visitor effect on zoo animals: Implications and opportunities for zoo animal welfare. *Animals : An Open Access Journal from MDPI, 9*, 366.

Sherwen, S. L., Harvey, T. J., Magrath, M. J. L., Butler, K. L., Fanson, K. V., & Paul H. Hemsworth. (2015). Effects of visual contact with zoo visitors on black-capped capuchin welfare. *Applied Animal Behaviour Science, 167*, 65–73.

Sherwen, S. L., Magrath, M. J. L., Butler, K. L., Phillips, C. J. C., & Hemsworth, P. H. (2014). A multi-enclosure study investigating the behavioural response of meerkats to zoo visitors. *Applied Animal Behaviour Science, 156*, 70–77.

Spruin, L., & Mozova, K. (2018). Specially trained dogs in the UK criminal justice system. *Seen and Heard, 28*, 63–67.

Stockham, R. A., Slavin, D. L., & Kift, W. (2004). Specialized use of human scent in criminal investigations. *Forensic Science Communications, 6*(3).

Suma, P., La Pergola, A., Longo, S., & Soroker, V. (2014). The use of sniffing dogs for the detection of Rhynchophorus Ferrugineus. *Phytoparasitica, 42*, 269–274.

Szokalski, M. S., Litchfield, C. A., & Foster, W. K. (2012). Enrichment for captive tigers (*Panthera tigris*): Current knowledge and future directions. *Applied Animal Behaviour Science, 139*, 1–9.

Szokalski, M. S., Litchfield, C. A., & Foster, W. K. (2013). What can zookeepers tell us about interacting with big cats in captivity? *Zoo Biology, 32*, 142–151.

Tait, P., & Farrell, R. (2010). Protests and circus geographies: Exotic animals with Edgley's in Australia, *Journal of Australian Studies, 34*, 225–239.

Tiira, K., Viitala, N., Turunen, T., & Salonen, T. (2019). Accelerant detection canines' ability to detect ignitable liquids. *Reviews of the Police University College of Finland*, 14/2019.

Tomkins, L. M., Thomson, P. C., & McGreevy, P. D. (2011). Behavioral and physiological predictors of guide dog success. *Journal of Veterinary Behavior, 6*, 178–187.

Troisi, C. A., Mills, D. S., Wilkinson, A., & Zulch, H. E. (2019). Behavioral and cognitive factors that affect the success of scent detection dogs. *Comparative Cognition & Behavior Reviews, 14*, 51–76.

Trousselard, M., Jean, A., Beiger, F., Marchandot, F., Davoust, B., & Canini, F. (2014). The role of an animal-mascot in the psychological adjustment of soldiers exposed to combat stress. *Psychology, 5*, 1821–1836.

Van De Werfhorst, L. C., Murray, J. L., Reynolds, S., Reynolds, K., & Holden, P. A. (2014). Canine scent detection and microbial source tracking of human waste contamination in storm drains. *Water Environment Research, 86*, 550–558.

van Dijk, B. (2018). Human rights in war: On the entangled foundations of the 1949 Geneva Conventions. *American Journal of International Law, 112*, 553–582.

Verhagen, R., Cox, C., Machangu, R., Weetjens, B., & Billet, M. (2003). Preliminary results on the use of Cricetomys rats as indicators of buried explosives in field conditions. In I. G. McLean (Ed.), *Mine detection dogs: Training, operations and odour detection* (pp. 175–193). Geneva: Geneva International Centre for Humanitarian Demining.

Ward, S. J., & Melfi, V. (2015). Keeper-animal interactions: Differences between the behaviour of zoo animals affect stockmanship. *PLoS One, 10*, e0140237.

Wark, J. D., Cronin, K. A., Niemann, T., Shender, M. A., Horrigan, A., Kao, A., & Ross, M. R. (2019). Monitoring the behavior and habitat use of animals to enhance welfare using the ZooMonitor app. *Animal Behavior and Cognition, 6*, 158–167.

Wharton, J., Khalil, K., Fyfe, C., & Young, A. (2019). Effective practices for fostering empathy towards marine life. In G. Fauville, D. L. Payne, M. E. Marrero, A. Lantz-Andersson, & F. Crouch (Eds.), *Exemplary practices in marine science education. A resource for practitioners and researchers* (pp. 157–168). Berlin, Heidelberg: Springer.

Wiggett-Barnard, C., & Steel, H. (2008). The experience of owning a guide dog. *Disability and Rehabilitation, 30*(14), 1014–1026.

Wilson III, C. (2018). Sacrificial structures: Waste, animals, and the monumental impulse. *Resilience: A Journal of the Environmental Humanities, 6*, 86–101.

Winkle, M., Crowe, T. K., & Hendrix, I. (2012). Service dogs and people with physical disabilities partnerships: A systematic review. *Occupational Therapy International, 19*, 54–66.

Young, A., Khalil, K. A., & Wharton, J. (2018). Empathy for animals: A review of the existing literature. *Curator: The Museum Journal, 61*, 327–343.

7 Animal healers

As discussed in Chapter 2 there are physical, psychological, and social benefits associated with pet ownership: this chapter considers efforts to harness these benefits within a therapeutic context. As Kruger and Serpell (2006) note, *Animal Assisted Therapies* (*AAT*), also variously referred to as *Animal Assisted Interventions* (*AAI*), form a sprawling literature with wide range of definitions and meanings given to the term. They cite the helpful definition from the Delta Society, an American organisation engaged in the certification of therapy animals:

> AAT is a goal-directed intervention in which an animal that meets specific criteria is an integral part of the treatment process. AAT is directed and/or delivered by a health/human service professional with specialized expertise and within the scope of practice of his/her profession. Key features include specified goals and objectives for each individual and measured progress (p. 23).

AAT has been widely used in many areas of practice (Fine, 2015), while inevitably there is some degree of overlap between categories, the application of AAT may be divided into the three broad areas of *mental health*, *physical health*, and *social wellbeing*. These areas will be considered before addressing the critical question of whether AAT works. Finally, a distinction may be drawn between *therapy animals* and *service animals*: the former are engaged in the therapeutic process, the latter, such as guide dogs, give assistance in everyday tasks.

AAT: Mental health

The introduction of animals, principally birds, dogs, dolphins and horses, into the treatment of mental health problems is predicated on the belief that this will increase the efficacy of the treatment and reduce the negative aspects of the condition and improve the patient's quality of life. AAT must be fully compliant with the ethical and legal principles which protect the rights of the individual (Koukourikos, Georgopoulou, Kourkouta, & Tsaloglidou, 2019; Yamamoto, & Hart, 2019).

Animals may assist in two ways with the therapeutic process: first, as an adjunct to therapy; second, as the focus of therapy. An example of the first role is provided by Lang et al. (2010) who used dogs to calm patients with schizophrenia during an assessment interview. In the second role the AAT may be either for mixed groups of psychiatric patients or for those with a specific type of disorder. Berget, Ekeberg, and Braastad (2008) evaluated the use of farm animals in AAT with patients with a range of psychiatric disorders including affective disorders, anxiety, personality disorders and schizophrenia. They found positive results but the design does not allow an understanding of the effect of the intervention according to diagnosis. However, there are studies of AAT with specific psychiatric conditions which allow more exact conclusions to be drawn.

Addictions

A cluster of studies have considered the use of horses in the treatment of substance use. Kern-Godal, Brenna, Kogstad, Arnevik, and Ravndal (2016a) give several reasons to introduce the horse into the therapeutic relationship. The inherent characteristics of the horse provide a model for cooperative behaviour allowing the patient to experience new emotions and behaviour. The more insightful patients may come to see the horse as a metaphor: it is non-judgmental, motivational and can build confidence and self-esteem; finally, interactions with the horse can illuminate human interaction.

In a Norwegian study Kern-Godal, Brenna, Arnevik, and Ravndal (2016b) interviewed eight patients, four males and four females aged between 20 and 30 years, who were taking part in a substance abuse treatment programme which included horse assisted therapy. The interviews revealed that patients saw the horses as facilitators of positive personal change and providers of emotional support during treatment. Kern-Godal et al. (2015) found that horse assisted therapy encouraged young people to engage in and complete treatment, although this completion effect was not found in a replication study (Gatti, Walderhaug, Kern-Godal, Lysell, & Arnevik, 2020). Kern-Godal et al. (2016b) describe how the stable (!!) environment of horse assisted therapy is viewed by patients as a break from treatment as usual while facilitating their personal change. Klemetsen and Lindstrom (2017) reported a systematic review of 10 evaluations of AAT, including both qualitative and quantitative studies, in treating substance use disorders. The evidence furnished by these studies was promising but Klemetsen and Lindstrom note that as most studies were exploratory a causal relationship between intervention and outcome could not be established.

There is a strong overlap between substance use and criminal behaviour and prison populations typically have a high number of addicts. There are several studies of the use of dog assisted treatment programmes for imprisoned addicts. For example, Contalbrigo et al. (2017) describe dog assisted therapy for offenders in prison in Padua, North-Eastern Italy. The sessions were focused on Maslow's hierarchy of need, beginning with physiological needs and progressing to various levels of psychological change. During the sessions the dogs were free to interact with the prisoners and parallels were drawn between canine and human needs. Contalbrigo et al. concluded that the treatment, in which the dogs played an integral role, improved the prisoners' psychological functioning while reducing some dysfunctional symptoms. Duindam, Asscher, Hoeve, Stams, and Creemers (2020) reported a meta-analysis of 11 evaluations of prison-based dog programmes finding a positive significant effect on both the prisoners' social-emotional functioning and their criminal recidivism.

Anxiety

Anxiety varies in intensity from a situationally-specific anxious state as, say, experienced by students waiting to go into the examination hall, to debilitating panic attacks, and to enduring states of anxiety after experiencing a traumatic event. The treatment of anxiety disorders is a major clinical enterprise in which animals may play a part. At the lower end of continuum there is some evidence that watching fish in an aquarium can bring about a sense of relaxation and lower anxiety (Clements et al., 2019). This observation may explain why aquariums are sometimes seen in waiting rooms: nothing soothes away the sound of the dentist's drill like a shoal of guppies. Indeed, some people have a genuine fear of visiting the dentist which is clearly not in their best interests. As well as aquatic life in the waiting room, a dog for the patient to stroke at their appointment may help reduce anxiety and ease dental treatment (Cruz-Fierro, Vanegas-Farfano, & González-Ramírez, 2019).

The experience of a traumatic event can leave a long-lasting psychological impact with a cluster of symptoms associated with the diagnosis of post-traumatic stress disorder (PTSD; see Chapter 3). The common causes of PTSD are accidents and the experience, as a participant or a victim, of violence. The growing realisation that some intervention was needed for soldiers returning from conflict experiencing severe anxiety akin to PTSD prompted the use of AAT. Thus, building on the extensive use of ATT in the treatment of trauma (Trzmiel, Purandare, Michalak, Zasadzka, & Pawlaczyk, 2019) treatments were developed, some with the involvement of animals, specifically for combat veterans (Lanning, & Krenek, 2013; Mims & Waddell, 2016). While the focus here is on AAT there are, in keeping with the distinction noted above, service dogs for veterans with PTSD (e.g., van Houtert, Endenburg, Wijnker, Rodenburg, & Vermetten, 2018).

Johnson et al. (2018) evaluated the effects of therapeutic horse riding on 29 veterans with PSTD. The veterans were randomly assigned to a horse-riding group or to a waiting-list control group. Alongside demographic and health history, PTSD symptoms were assessed with a standardised checklist; other measures of psychological functioning, such as coping self-efficacy and difficulties in emotional control, were also included. In the sessions the veterans learned the basics of horse care and horsemanship skills and they completed supervised tasks on horseback. After both 3 and 6 weeks of therapy there was a significant decrease in scores on the PTSD scale as compared to the control group. The other psychological measures did not show significant change. Arnon et al. (in press) reported preliminary data on a similar programme for veterans with PTSD. They found that clinician-assessed PTSD decreased from pre- to post-treatment and at a 3-month follow-up; the same pattern was evident for measures of anxiety and depression.

Boss, Branson, Hagan, and Krause-Parello (2019) included nine studies in a systematic review of equine-assisted interventions with military veterans. They noted mixed outcomes, some positive others not, for effects on PTSD symptoms but little effectiveness on other measures of psychological and social functioning. They conclude that equine assisted intervention for PTSD in military veterans "May be effective, however, we cannot make a definitive determination based on the current review of evidence" (p. 30).

Autism and autism spectrum disorder

The condition known as autism has had several diagnostic revisions (it follows that over time researchers will use changing diagnoses). In 1994 the American Psychiatric Association (APA; DSM-IV) categorised autism as a spectrum ranging from mild to severe. To aid diagnosis, DSM-5 (APA, 2013) introduced the term *autism spectrum disorder* defined by two groups of symptoms, each consisting of specific behaviours, evident in early childhood: (1) persistent impairment in reciprocal social communication and social interaction; (2) restricted, repetitive patterns of behaviour. The psychological and social consequences of ASD are felt by both the individual and their family. Animals, principally dogs and horses but also dolphins, guinea pigs, llamas and rabbits (e.g., Griffioen, van der Steen, Cox, Verheggen, & Enders-Slegers, 2019; O'Haire, McKenzie, Beck, & Slaughter, 2019) have been incorporated into a plethora of different therapies. Several illustrative studies are described below.

Can a pet dog go some way towards alleviating the difficulties faced by children with ASD and their families?

Wright et al. (2015a) looked at the effect of owning a dog on family functioning – using a standardised scale to assess family strengths and weaknesses at pre- and post-ownership and at follow-up – by comparing parents of children with ASD who had recently acquired a pet dog with a control group without a dog. A parental-report measure of child anxiety was completed by a sub-group of families. There was a significant improvement in family functioning in the dog owing group compared with the control, while the anxiety levels in the dog-owning group fell more than in the controls. Wright et al. (2015b) considered the effect of a acquiring a pet dog on the stress levels of 38 primary carers of children with ASD compared with 24 controls without a dog. Stress levels were assessed at pre- and post-intervention and at follow-up using a standardised measure of parenting stress. There was a significant improvement in stress levels for the dog acquisition group compared to the controls, with some carers moving from clinically high to normal levels on the parental distress subscale. Hall, Wright, Hames, Team, and Mills (2016) conducted a follow-up of over 2 years with some of the families who had taken part in the two Wright et al. studies and found that the beneficial effects of dog ownership were maintained over time.

A Dutch randomised control trial by Wijker, Leontjevas, Spek, and Enders-Slegers (2020) looked at the effects of AAT with dogs on 53 adults aged between 18 and 60 years with ASD. The intervention successfully reduced perceived stress and symptoms of agoraphobia and improved social awareness and communication.

Moving on to horses, Ward et al. (2013) explain the rationale for therapeutic riding (TR) for those with ASD:

> Therapeutic riding emphasizes control, attention and focus, sensory management, and communication (verbal and/or nonverbal) in order to teach riding skills. Furthermore, TR provides a multisensory experience. Contact with animals, including horses, stimulates physiological, psychological, and social responses in children and adolescents. … Consequently, TR may be particularly effective for children with autism spectrum disorders (ASD) who experience difficulties with joint attention, appropriate social responses, communication, and management of sensory input and responses (p. 2190).

Ward et al. evaluated a 30-week TR programme for 21 autistic children, mean age 8.1 years. Each session was based on the four themes of *orientation*, such as preparing food for the horses, followed by *mounting and riding, riding skills,* and finally *closure* where the children said goodbye their horse. The teacher ratings of the children showed a significant increase in social interaction, improved sensory processing, and a decrease in symptom severity. Gabriels et al. (2015) carried out a randomised control trial of TR with 116 children aged between 6 and 16 years with a diagnosis of ASD. The TR group showed significant positive changes in social cognition, social communication and language.

Anderson and Meints (2016) report a study involving children aged from 5 to 16 years with a diagnosis of ASD. The children engaged in a 5-week programme with a 3-hour session per week addressing therapeutic riding, horsemanship (grooming etc.) and stable management (cleaning, feeding etc.). The findings showed an overall reduction in maladaptive behaviour but no significant change in social and communication skills.

A South Korean study reported by Kwon, Sung, Ko, and Kim (2019) looked at the effects of TR in addition to conventional therapy on the language and cognitive functioning of children aged between 6 and 11 years with ASD or intellectual disability. The TR was similar to that in the Anderson and Meints (2016) study. As compared to a control group using conventional therapy, those children also participating in the TR showed improvements in language and cognition. However, 8 weeks after treatment the differences between the groups had disappeared.

There have been several systematic reviews of AAT with ASD. In a review of 14 studies, O'Haire (2013) noted that the most common targets for change were social interaction, language and communication, positive emotional experience, and motor skills. O'Haire points to several limitations across the evidence base including an absence of standardised, replicable treatments, weak research designs, and varying outcome measures. O'Haire concludes that: "Current practices should be viewed as potentially promising enrichment interventions, rather than stand alone or complementary evidence-based treatments" (p. 213).

In her second systematic review O'Haire (2017) included 28 studies reported between 2012 and 2015. As before, dogs and horses were the animals most frequently and the treatment targets remained relatively constant, with social interaction most often reported as showing a significant improvement. This review concludes that: "Based on the existing evidence from 28 studies synthesized in this systematic review, the provision of AAI for autism should be viewed as a possibly efficacious enrichment activity for autism that may increase social interaction" (p. 212). As have others, O'Haire cautions that the practice of AAT should be seen as having the potential to enrich interventions while *not* an evidence-based treatment in and of itself. Alexandra et al. (2019) reported a systematic review of seven evaluations of equine assisted activities and therapies for children with ASD. They concluded that horses could be a valuable asset in therapy but that definitive evidence is awaited. Trzmiel, Purandare, Michalak, Zasadzka, and Pawlaczyk (2019) conducted a meta-analysis of 15 evaluations of equine assisted activities and therapies in children aged between 3 and 16 years diagnosed with ASD. The analysis indicated that equine-based activities could be beneficial, particularly with regard to social functioning, but that substantial variations across studies make it impossible to draw definite conclusions. Tan and Simmonds (2019) reached a similar conclusion from their review of 16 studies of equine assisted interventions with children with ASD. They note the benefits, particularly greater social interaction and fewer problem behaviours, but point to variability in the presentation of the intervention across studies. They suggest TR may be particularly useful when there has been limited success with traditional clinic-based types of treatment.

As well as ASD there are other childhood and adolescent disorders where animals are part of the therapy.

Childhood and adolescent disorders

The issues here fall into two categories: first, disorders defined by this developmental period (some of which may continue in later life) such as childhood phobias; second, mental health problems, such as anxiety and schizophrenia, evident in all age groups.

"Developmentally-defined" disorders

Attention deficit hyperactivity disorder (ADHD) is the most prevalent neurodevelopmental disorder, found in 3 to 7% of school aged children. ADHD is characterised by age-inappropriate levels of inattention and/or hyperactivity and impulsive behaviour. ADHD creates difficulties in social relationships and impairs the child's education, while many children also have problems with balance and physical coordination. Given the range of symptomatology a variety of treatments, including pharmacological, behavioural and cognitive-behavioural, neurofeedback, and physical exercise, have been used with ADHD. Busch et al. (2016) suggest that the beneficial effects such as calming, motivation, and socialisation of animal-assisted interventions make ADHD a candidate for this approach. In particular, the horse has been widely used with children with ADHD.

In a typical study Jang et al. (2015) evaluated equine assisted therapy with 20 children aged 6–13 years diagnosed with ADHD. The content of the therapy was similar to that outlined above for children with ASD. The evaluation showed an improvement in the core symptoms for 18 of the children which may be due to the beneficial effects of physical exercise on attention, cognition and motivation. In addition, the children learned emotional control which may have a positive influence on social behaviour.

Child abuse can take the form of neglect, psychical abuse and sexual abuse; some children experience more than one type of abuse. The physical, psychological and social consequences of childhood abuse are profound in both the short- and long-term and a great deal of therapeutic endeavour, some including animals, has been devoted to working with abused young people (Narang, Schwannauer, Quayle, & Chouliara, 2019). The start of treatment may be an interview with the child to discover more about the allegations of abuse and the disclosure of sexual abuse can be extremely stressful for the child. Krause-Parello, Thames, Ray, and Kolassa (2018) interviewed 51 children aged from 4 to 16 years following an allegation of sexual abuse looking at whether the presence of a dog could lower the children's stress during the interview. The children were randomly allocated to an interview with the dog present which the children could stroke or to an interview with no dog. Alongside the child's self-report, several biomarkers of stress, such as heart rate and blood pressure, were collected pre- and post-interview. It was found that the presence of a dog significantly lowered the children's stress levels.

An Australian programme reported by Kemp, Signal, Botros, Taylor, and Prentice (2014) taught basic horse riding and management to 15 children and 15 adolescents, both Indigenous and non-Indigenous, who had been sexually abused. The teaching exercises, focused on anxiety and trauma, were designed "To address issues such as: trust, communication, boundaries, observation, body language, attitude and self-perception" (p. 561). Kemp et al. found that the programme had significant beneficial effects upon the young people's psychological functioning regardless of their ethnicity. Signal, Taylor, Botros, Prentice, and Lazarus (2013) used a similar programme to Kemp et al. but aimed at reducing symptoms of depression in 44 survivors, 10 of whom were Indigenous Australians, of sexual abuse. Those taking part in the programme were 15 children (aged 8–11 years), 15 adolescents (aged 12–17 years), and 14 adults (aged 19–50 years). The evaluation used an age-appropriate measure of depression (Child Depression Index or Beck Depression Inventory) administered at intake to service, post in-clinic counselling, and post-AAT. It was found that, regardless of age or ethnicity, AAT resulted in a reduction in scores on the depression inventory.

Dietz, Davis, and Pennings (2012) used AAT in group therapy with sexually abused children. They used three conditions – treatment with no dogs, treatment with dogs and therapeutic stories, treatment with dogs and without therapeutic stories – incorporating the three elements of group therapy, a dog, and therapeutic stories involving the dog. It was found that the children's scores for anger, anxiety, depression, dissociation, PTSD and sexual concerns fell significantly from pre- to post-test for both the dogs no stories and the dogs with stories groups. An Australian programme reported by Signal, Taylor, Prentice, McDade, and Burke (2017) used AAT to treat PTSD symptomology in sexually abused children. A group of 20 children, both Indigenous and non-Indigenous, composed of 12 males and 8 females aged from 5 to 12 years took part. In the first 3 weeks of the programme the groups of 4 to 6 children visited the local RSPCA shelter where they interacted with a trained therapy dog and its handler. These sessions had a specific therapeutic focus, such as recognising cruelty to animals, and were delivered by a specialist RSPCA education officer. In the final 7 weeks the children along with their social workers applied the learning from their time with the dogs to aspects of human interaction such as body language and feelings, managing emotions, developing and respecting boundaries, and requesting support when needed. As measured by assessment of PTSD symptomatology, the treatment was effective for both males and females and the Indigenous and non-Indigenous children.

Mental health problems

AAT, primarily with dogs and horses, has been used in the treatment of child and adolescent mental health problems. Stefanini, Martino, Allori, Galeotti, and Tani (2015) considered the effects of AAT with a sample of 34 children and adolescents, aged from 11 to 17 years, with a psychiatric diagnosis. The young people were randomly assigned to treatment and control groups and over a 3-month period the intervention was delivered to 17 young people as a structured weekly session. These young people first became familiar with the dog and its handler, then participated and individual and group activity, and finally reviewed the experience of AAT in which they had interacted with a dog by playing, petting, grooming, walking and giving basic commands. The outcome showed that those in the treatment condition showed improvements in global functioning and school attendance, alongside a significant reduction of time spent in hospital. At a 3-month follow-up the treatment group showed a significant improvement in social activity and social skills and additional, more affectionate, interactions with their assigned animal.

Stefanini, Martino, Bacci, and Tani (2016) looked at the effect of AAT on the behavioural and emotional functioning of 40 child and adolescent psychiatric patients who were hospitalised for acute mental disorders. Equal numbers of young people were randomly assigned to the treatment and control conditions: there were 10 treatment sessions comprising 5 group therapy and 5 individual therapy. All the sessions, similar to those in the Stefanini et al. (2015) study, included a dog its handler. The treatment group showed significant improvements in emotional and behavioural symptoms which were maintained at a 3-month follow-up.

Hoagwood, Acri, Morrissey, and Peth-Pierce (2017) carried out a systematic review of 24 experimental studies published between 2000–2015 of AAT for children or adolescents with mental health conditions. The majority of studies were of equine therapy for autism. They concluded that while the evidence base is sparse, equine therapy for autism and canine therapy for childhood trauma are generally promising.

Jones, Rice, and Cotton (2019) reported a systematic review of seven studies in which canines were incorporated into mental health treatments for the 10 to 19-year age group. They found that the interventions were varied and that details of the role of the dogs was lacking. However, the dogs did bring benefits on symptomatology across several disorders including PTSD, anxiety, and anger dysfunction.

AAT: Physical health

Animals may play a role in the detection and treatment of physical ailments. There are many examples of this application of ATT to be found across a range of conditions as diverse as burns (Pruskowski, Gurney, & Cancio, 2020), sleep disturbance (Koskinen et al., 2019), malaria (Kasstan et al., 2019), and palliative care (Engelman, 2013). The focus here is on four areas – cancer, epilepsy and seizures, heart disease and pain – where bodies of research have developed that highlight this form of ATT and its attendant issues.

Cancer

Given the importance of early detection of cancer our pets may have a role to play in saving lives (Roncati, 2019). This proposition requires empirical support prior to being used to inform medical practice and has attracted a body of research. Buszewski et al. (2012) suggest that dogs may be of use in screening for cancer as the detection of the odours of certain volatile organic compounds (VOCs), present in breath and urine, act as detectable markers. Willis et al. (2004) trained dogs to recognise the odour in urine caused by bladder cancer. In a test with 36 samples they found that the dogs made more successful discriminations than would be expected simply on the basis of chance. Pickel et al. (2004) found promising results with dogs trained to detect melanoma, while Cornu, Cancel-Tassin, Ondet, Girardet, and Cussenot (2011) and Taverna et al. (2015) both reported that trained dogs had a high rate of identifying prostate cancer from urine samples. A Mexican study by Guerrero-Flores et al. (2017) found that a trained beagle showed high levels of detection of cervical cancer from a variety of odour-bearing sources.

McCulloch et al. (2006) successfully trained "ordinary household dogs" to discriminate between breath samples from controls and both lung and breast cancer patients. Albertini, Mazzola, Sincovich, and Pirrone (2016) considered the use of trained dogs to detect human lung cancer VOCs in urine. They allocated 150 participants to one of three groups giving 57 patients with lung cancer, 38 patients with a non-cancerous lung disease, and 55 healthy controls. The dogs had a mean detection rate of over 80%. Albertini et al. make the point that not only can the dogs discriminate between lung cancer and healthy patients but between cancer and other lung diseases. A similar study by Montes et al. (2017) replicated the Albertini et al. study, as did Mazzola et al. (2020), with urine samples rather then breath as the sample material to test for lung cancer.

As well as aiding detection, animals can be a source of support for those receiving cancer treatment. White et al. (2015) asked women about their experience of therapy dogs in their counselling sessions. The women spoke of the benefits of the ATT in terms of successful initiation, engagement and personal disclosure in the sessions.

Epilepsy and seizures

As Bishop and Allen (2007) explain, epilepsy is an umbrella term for a range of conditions associated with different types of seizure which may cause physical convulsions and loss of consciousness. These seizures are associated with a range of psychological and social difficulties and there several treatments ranging from the pharmacological to the psychotherapeutic. The detection of the onset of a seizure has several advantages including the possibility of a reduction in the frequency of seizures: Brown and Strong (2001) make the observation that:

> Some doctors who treat epilepsy (and who listen to what their patients tell them), will have heard anecdotal reports of pet dogs reported to display some sort of premonitory behaviour before their significant human has a seizure. This behaviour is often reported to occur when neither the person with epilepsy nor other family members share the premonition (p. 39).

There is some support for the efficacy of seizure alert dogs generally (Brown, & Goldstein, 2011; Brown & Strong, 2001; Strong, Brown, Huytens, & Coyle, 2002) and with children specifically (Kirton, Wirrell, Zhang, & Hamiwka, 2004). These findings suggest that refinements in training could increase the frequency of successful detection. The starting point for such training is specification of exactly what the dogs detect: is it a change in odour or subtle changes in behaviour? Catala, Cousillas, Hausberger, and Grandgeorge (2018) reviewed 28 studies of dogs which are alert to different types of seizure. The dogs varied in size, breed and training alongside using a range of altering cues including scent and behaviour. The studies indicated reasonable detection rate and a reduction in the frequency of seizures and reported increased quality of life. Catala et al. state that across studies there was a generally low level of a reasonable methodological rigour so that any conclusions regarding the use of seizure dogs must remain tentative.

The point raised by Catala et al. (2018) regarding the type of alerting cue is important as its precise identification would be invaluable for training purposes. An experimental study by Catala et al. (2019) demonstrated that odour was the discriminative cue to which the seizure alert dogs responded. This view was strengthened by van Dartel et al. (2020) who used electronic methods to detect a VOC for epilepsy with a reasonably high degree of accuracy.

Heart disease

Heart disease has three phases: (i) the presence of risk factors; (ii) the onset of the illness that requires medical intervention; (iii) the aftermath. Animals may play a positive role at all three phases. Cardiovascular disease (CVD) can occur when fatty deposits accumulate and cause narrowing of the coronary arteries: alongside hypertension, CVD is the most common cause of acute myocardial infarction (AMI) or heart attack. Some of the risk factors for heart disease can be brought under control by lifestyle changes such as diet and physical exercise. As discussed in Chapter 2 and as evident in the large-scale surveys from Europe (Maugeri et al., 2019) and the USA (Krittanawong et al., 2020) dog ownership can have benefits for physical and psychological health by reducing stress and encouraging exercise.

Risk factors for CHD

The American Heart Association conducted a review of the evidence addressing pet ownership and risk of CVD (Levine et al., 2013). They make the point that a pet is not a guarantee of prevention, rather a pet may be a conduit to success:

> The writing group emphasizes that although pet adoption, rescue, or purchase may be associated with some future reduction in CVD, the primary purpose of adopting, rescuing, or purchasing a pet should not be to achieve a reduction in CVD risk. Furthermore, the mere adoption, rescue, or purchase of a pet, without a plan of regular aerobic activity (such as walking a dog) and implementation of other primary and secondary cardiovascular preventive measures, is not a sound or advisable strategy for reduction in CVD risk (p. 2360).

Friedmann, Thomas, Son, Chapa, and McCune (2013) considered the effects of pet ownership on reducing hypertension in a sample of mainly women aged from 50 to 83 years. They reported that the presence of a pet cat or dog had a significant positive effect in lowering the blood pressure of those who took medication for hypertension or had blood pressure in the pre- to mild-hypertensive range. Krause-Parello and Kolassa (2016) monitored the effects of a visits from a pet therapy dog and handler on the blood pressure and heart rate of 28 female and males, aged from 60 to 102 years, living in the community in New Jersey. In a crossover design there were two home visits, about a week apart: first a therapy session with a handler-canine pair, second a visit from a volunteer without a therapy dog. The pet therapy significantly lowered both blood pressure and heart rate.

A Swedish study by Handlin, Nilsson, Lidfors, Petersson, and Uvnäs-Moberg (2018) considered the effects of a therapy dog on the blood pressure and heart rate of older nursing home residents aged from 70 to 100 years. A dog and handler visited residents at three homes while in the control condition two researchers visited three different homes. When the dog visited, she placed her head on the person's lap while making eye contact which, guided by the handler, led the elderly person to stroke and interact with dog. There were significant decreases in both systolic blood pressure and heart rate during the therapy dog visits. Handlin et al. note that as the visit was by a dog–handler team the beneficial effects may be due to this combination rather than just the dog. Why the dog visits had a positive effect is not clear: Handlin et al. speculate that the benefits may be mediated by changes in autonomic nervous system activity.

Schreiner (2016) included CVD risk factors in an overview of the effects of pets on CVD prevention, noting that walking with dogs is a relatively consistent benefit with regard to cardiovascular risk. However, Schreiner makes the point that while there is a connection between dog ownership and walking, the heath benefits are not universal. This fact implies that other factors not connected to the dog, such as socioeconomic status or a strong bond with the pet, may be important in understanding the overall picture. A study conducted in the Czech Republic by Maugeri et al. (2019) addressed the issue of lifestyle, dog walking and cardiovascular health. A total of 1769 females and males, aged between 25 and 64 years, without a history of CVD took part in the study.

The focus was on three groups, pet owners generally, dog owners specifically, and non-pet owners. These groups were compared on a range of variables including CVD risk factors and cardiovascular health (CVH) metrics such as blood pressure, body mass index, cholesterol, diet, physical exercise, and smoking. Within the sample there were 746 pet owners, of whom 429 owned a dog, and 1,023 did not have a pet. The pet owners generally, and dog owners specifically, were more likely to report physical activity, diet, and ideal levels of blood glucose but a *high* level of smoking: overall, the pet owners had a higher CVH score than non pet owners. When adjustment for covariates such as age and sex were made, dog owners exhibited higher CVH scores than non-pet owners' other pet-owners, and non-dog owners. A Swedish population study by Mubanga et al. (2019) identified a cohort of 2,026,865 individuals from the Register of the Total Population specifically recording data on a range of variables including dog ownership, education level, hospital admissions and prescribed medication. The cohort was followed for 6 years noting whether an individual took medication for a cardiovascular risk factor. The results show that dog owners have a marginally *higher* chance of starting anti-hypertensive and lipid-lowering treatment than non-owners, particularly among those aged 45–60 years. This finding suggests that a lower risk of CHD evident in previous studies is not explained by reduced hypertension and dyslipidaemia (abnormal amount of lipids such as cholesterol in the blood). However, Mubanga et al. point to limitations such as the potentially confounding effects of unmeasured variables such as socio-economic factors, health status before acquiring a pet, primary pet responsibility, and physical activity related to dog walking.

The picture that emerges is one of uncertainty: there is no doubt that exercise is to be recommended and for some owners a dog is associated with increased levels of activity. Yet further, a dog's presence can also be relaxing and act to lower heart rate and blood pressure (Machová, Poběrežský, Svobodová, & Vařeková, 2017) for children as well as adults (Xu et al., 2017). However, unravelling precisely the psychological, medical and social variables which, if at all, link pet ownership and amelioration of the risk of CHD as yet remains elusive.

Treatment for CHD

There are obvious hygienic disadvantages to the presence of dogs in a hospital but there are also advantages to be had in terms of patient benefit (McCullough, Ruehrdanz, & Jenkins, 2016). With regard to CHD Cole, Gawlinski, Steers, and Kotlerman (2007) considered the effects of visits from a therapy dog on patients hospitalised with heart failure. They found that the AAT was of benefit in improving cardiopulmonary pressures, neurohormone levels, and reducing anxiety. Stefanini, Martino, Bacci, and Tani (2016) carried out a small-scale AAT study with 11 patients in hospital awaiting a heart transplant. They reported that patients welcomed the visits from the dog volunteer handler and in some cases found compensation for missing their own dog.

Recovery from CHD

Friedmann, Katcher, Lynch, and Thomas (1980) examined the survival rates of 92 patients, 29 women and 67 men, who received hospital treatment for myocardial infarction (48 patients) or angina pectoris (44 patients). The 1-year survival rate showed that 78 of the 92 patients were alive 1 year after their hospital admission. It was known that 53 of the 92 patients had 1 or more pets. There was a significant relationship between pet ownership and 1-year survival status such that of the 11 of the 39 patients who did not own pets died, compared to 3 of the 53 pet owners.

Further analysis incorporating a range of social variables led Friedmann et al. to conclude that:

> The beneficial effect of pet ownership is not a statistical artifact produced by differences in age or health status between patients with and without pets. Moreover, the benefit is probably not a result of the protective effect of the physical activity needed to walk dogs, since owners of pets other than dogs had a better survival rate than the subjects without pets. Currently, the major unanswered question relates to the source of the apparent influence of pets on survival (p. 310)

The Swedish population study, noted above, by Mubanga et al. (2017) found that dog owners were at a lower risk of death than non-owners after a heart attack and after an ischemic stroke. Machová, Procházková, Říha, and Svobodová (2019) found that when AAT (with a dog) was added to standard physiotherapy and occupational therapy for stroke patients there was no effect on blood pressure and heart rate. However, when asked to rate how they felt, the patients with the added AAT were significantly more likely to say that they felt better after the AAT sessions suggesting that AAT may be of psychological benefit.

Yeh, Lei, Liu, and Chien (2019) conducted a systematic review and meta-analysis of 12 studies of the association between pet ownership and cardiovascular disease (CVD). They reported there was not an association between pet ownership and "All-cause mortality or the CV outcomes of CVD risk, adjusted CV mortality, or risk of MI or stroke. Despite this, subgroup analysis showed an association between pet ownership and was associated with a lower CVD mortality in the general population, and between pet ownership and adjusted CVD risk in patients with established CVD" (p. 12).

Kramer, Mehmood, and Suen (2019) conducted a systematic review and meta-analysis of ten studies of the association between dog ownership with all-cause mortality, death with and without prior cardiovascular disease and cardiovascular mortality. They concluded that "Dog ownership is associated with reduced all-cause mortality possibly driven by a reduction in cardiovascular mortality" (p. 7).

Pain

We are all familiar with the experience of acute pain but for some people pain is chronic, a part of their everyday life. Marcus et al. (2012) looked at the effect of therapy dogs in an adult outpatient pain management clinic. They found that the dog's presence was associated with a significant reduction in pain and emotional distress for the patients and their accompanying relatives and friends. Marcus et al. (2013) used AAT in an outpatient setting for patients with fibromyalgia, a condition where the individual may experience continuous pain throughout their body but particularly so in, say, the back or neck. There were significant improvements in pain relief, mood, and other measures of distress for those patients whose treatment included a therapy dog.

Harper et al. (2015) considered the effect of AAT with patients who had undergone arthroplasty, a surgical procedure to restore the function of a joint. They reported that the therapy dogs had a positive effect on both the patients' levels of pain and their satisfaction with their stay in hospital. Carr, Wallace, Onyewuchi, Hellyer, and Kogan (2018) interviewed 12 Canadian patients, aged from 39 to 70 years, receiving treatment for long-term chronic pain. The patients, all living with a dog, answered questions about the role of their dog in their everyday life. The response analysis identified four themes: (1) in the absence of routines found with work and friends the dog gives meaning to life; (2) by responding to their needs and providing comfort the dog is a caregiver; (3) the dog gives emotional support with the pain is at its worst; (4) the presence and behaviour of the dog provides companionship. Thus, for some in chronic pain, a dog improves the quality of life, mentally, physically and socially, relieves suffering and gives a reason to focus on the future. Brown, Wang, and Carr (2018) added assisting sleep to the role of the dog in assisting those with chronic pain. Braun, Stangler, Narveson, and Pettingell (2009) used AAT (with dogs) to enhance pain relief with children aged 3 to 17 years in an acute care pediatric setting. They reported that for those children participating in AAT the reduction in pain was four times greater than for children relaxing for 15 minutes.

Carr, Norris, Hayden, Pater, and Wallace (2020) reviewed eight studies looking at dog ownership and health in people with chronic pain. The general conclusion is that there was little effect on pain but better mental health and social support was reported by those living with a dog. However, Carr et al. point to a lack of quality in some studies and that comparisons across studies are restricted by variations in outcome measures. Waite, Hamilton, and O'Brien (2018) reported a meta-analysis of the effects of AAI on pain, anxiety and distress in medical settings. They reported an overall positive effect of AAI for all three conditions generally while pointing to the need for improved methodology and greater uniformity across studies.

Finally, as the world moves on so new challenges arise. As I write this in the summer of 2020, the world is gripped by the COVID-19 pandemic, acutely so in Leicester where we are enduring an extended lockdown. In the global struggle the first steps have been taken to bring animals to the crisis. Preliminary findings made available by Grandjean et al. (2020) report on the use of dogs to detect the scent of COVID-19. They found that trained dogs had a very high rate of positive identification from armpit sweat samples of identifying people with the virus. Hoy-Gerlach, Rauktis, and Newhill (2020) make the case, extrapolating from what is known of human-animal companionship, that pets will offer support for the stressors such as social distancing brought about by the COVID-19 pandemic.

AAT: Social wellbeing

There are several aspects of our everyday lives that we rely on for our social wellbeing: we need a place to live, it is important that members of our community are helped in time of need, and we want to be safe from criminal behaviour. There are occasions when these social needs are not met and we call on animals to help alleviate the situation.

Homelessness

A person can become homeless for a varying period of time for several reasons including poverty, physical and mental ill-health, drug use, and family breakdown. The charity *Shelter* estimates that there are 320,000 homeless people in the UK; the Genesis trust (www.genesistrust.org.uk) cites figures estimating that between 5 to 10% of homeless people have pets, most frequently dogs. While this estimate is based on research from

North America the figure for the UK may fall into the same range. If this is so then based on the estimate from *Shelter*, then there are between 16,000 and 32,000 homeless animals on our streets. This is not, however, the problem it may appear: a dog does not need four walls and a roof, it does need food, warmth and companionship which all may be available regardless of the owner's homelessness [see Kim (2019) for an incisive discussion of this point].

An American study by Cronley, Strand, Patterson, and Gwaltney (2009) looked at the characteristics of 4,100 homeless people, 1,664 men and 2,436 women with a mean age of 39 years, who cared for animals. The characteristics of those more likely to care for an animal were Euro-American women who had experienced domestic violence and were homeless for the first time.

At a fundamental level, the reasons why homeless people have pets is no different to any other person: pets provide companionship and emotional support, promote a sense of responsibility and help combat loneliness (Ferrigno, 2015; Irvine, 2013; Rew, 2000). Howe and Easterbrook (2018) interviewed seven homeless people, two women and five men, in two English cities. Six of the seven lived in homeless accommodation and one slept rough: five people owned dogs, one person's dog had recently died and one owned a rat. The pet owners said that there were costs associated with their pet such as limited mobility and difficulties accessing services. However, these costs were outweighed by the benefits the pet provided such as companionship, a sense of responsibility and an increased resilience to substance abuse. The grief resulting from the loss of a pet could make interviewees turn to drugs as a way of coping with their feelings. However, given their situation there are specific reasons why homeless people may have a pet. Kerman, Gran-Ruaz, and Lem (2019) reviewed 18 studies concerned with pet ownership and

Figure 7.1 Homeless Man and His Dogs.
Source: Photograph by Nick Fewings on Unsplash.

homelessness finding three main reasons for pet ownership: (1) mental health and purpose; (2) social support and connection; (3) access to housing, employment, and services. The issues of physical health and experience of crime, particularly violent crime, were studied less often.

Mental health

Mental health problems can precipitate homelessness, while sleeping rough may exacerbate poor mental health. It is clear that alongside a range of physical ailments, mental heath problems such as depression, psychosis and addiction, are over-represented in homeless populations (e.g., Aldridge et al., 2018; Kidd, Gaetz, & O'Grady, 2017). It has been suggested that the continual, everyday negative experiences of the homeless person are related to mental heath problems generally and depression specifically (Fitzpatrick, Myrstol, & Miller, 2015). Is it possible that owing a pet could act as a buffer against poor mental health?

A Canadian study by Lem et al. (2016) compared two groups of homeless young people, aged between 16 and 24 years: there were 89 pet owners and 100 non-owners. When controlling for variables such as gender and regular drug use, pet ownership was *negatively* associated with depression. Indeed, the odds of depression were three times greater for the youth people without a pet. Cherner et al. (2018) reported that 50 of 501 homeless men and women living in emergency shelters in three Canadian cities had pets. However, there was no relationship between pet ownership and mental health. This null finding may be influenced by the source of the sample, sheltered accommodation, as the majority of such organisations do not allow pets (see below).

The possibility of violence, including sexual violence, is an ever-present risk for many homeless people (Heerde & Pallotta-Chiarolli, 2020; Nilsson, Nordentoft, Fazel, & Laursen, 2020; Wenzel, Koegel, & Gelberg, 2000). While violence may result in serious physical injury it also carries the risk of psychological harm such as severe anxiety (Welfare & Hollin, 2015) and exacerbating existing conditions such as PRSD (Taylor & Sharpe, 2008). A dog may provide some protection against violent attacks.

Social support

There are various sources of social support available to most people such as formal support from health services or informal such as friends and family. These social supports are important for our well-being. As the length of time an individual is homeless lengthens so their social networks will become increasingly fragmented (Green, Tucker, Golinelli, & Wenzel, 2013; Hwang et al., 2009). A reduction in social support increases the risk of negative consequences such as ill-health and criminal victimisation. However, as documented by Reitzes, Crimmins, Yarbrough, and Parker (2011), homeless people may form their own support structures and networks to provide mutual advice and social contact.

As many homeless people and their pets are visible on city streets so support or otherwise can come from members of the public. Taylor, Williams, and Gray (2004) asked 90 members of the public in Cambridge, England, for their opinions about homeless people who owned dogs. There was a majority view that homeless people should be allowed pets but with some concern expressed, mainly by women, for the dog's welfare. Irvine, Kahl, and Smith (2012) interviewed 60 homeless people, 32 women and 28 men, in Sacramento, California asking about their encounters with the public. They said the criticism they most often faced was that they were not in a position to own a dog. Their stock response is that they are "homeless not helpless" and can provide for and value their pet just as well as people with housing.

Access to services

There are a range of public services such as transport, medical and mental health services which homeless people may find it difficult to access. In order to try to remedy this situation there are organisations, some faith-based, specifically for homeless people. American studies by Donley and Wright (2012) and by Fitzpatrick et al. (2015) reported that homeless people may become resistant to these types of organisation. This resistance may be due to what Hoffman and Coffey (2008) term *objectification* and *infantilization*: i.e., not being seen a person but as a number or a child. Thus, some homeless people opt out of contact with some organisations in order to maintain their dignity and self-respect. However, homeless people must sometimes make a choice between their pet or a roof over their head. Howe and Easterbrook (2018) note that in London just 37 of the 222 homeless accommodation projects accept dogs and only three of the equivalent services in Manchester will allow dogs. Similar observations have been made by commentators in America (Rhoades, Winetrobe, & Rice, 2015) and Australia (Cleary et al., 2020). Scanlon, McBride, and Stavisky (in press) sent out an online survey to 523 UK homelessness service providers and received 117 responses. Of these 117, 43 provided services to pets although 90 said that they had had requests to accommodate a pet. The reasons most often given for accepting a pet were the benefit to the owner or to the animal. The majority of the services that allowed pets had policies for animal welfare safeguarding and damage restriction. In contrast, pets were not allowed by 74 organisations: these providers mainly cited the health and safety of staff and residents as a primary concern.

Scanlon, McBride, and Stavisky make the point that that the demand for pet-friendly accommodation for homeless people far outstrips what is available. Maharaj (2016) argues that a strong link between homeless people and their pet should be seen as an opportunity for an effort to be made to encourage and support service providers to accommodate pets, perhaps by having designated space for animals. This point is given added emphasis by a Canadian study which showed that while pets are particularly important to homeless vulnerable women, they either cannot use some services because of their pet or must surrender the animal (Labrecque & Walsh, 2011).

Surrogate homes

The occasion may arise when it is necessary for a person, typically because of a vulnerability, to live in a home that is not their own cared for by other people. Two common examples of this situation occur when a family breakdown leaves a child without a home and when older adults require residential care. What role might animals play in these settings?

Foster homes

Carr and Rockett (2017) considered eight young people, three boys and five girls, aged 10 to 16 years, living in long-term foster care. All the foster families had a pet dog. Seven of the young people had a background of neglect and abuse and presented with a range of difficulties including distress and angry outbursts. The children said that in their current strange situation the animal gave them a sense of being cared for and security. The children also said that in times of distress the dogs offered reassurance and comfort.

Care homes

The introduction of animals into care homes for elderly citizens, either on a permanent or visiting basis, was initially seen as a way of helping to improve the mood of the residents (e.g., Crowley-Robinson, Fenwick, & Blackshaw, 1996). While improving the well-being of care home residents remains an aim, the methods have become rather more sophisticated including specialist animal assisted interventions for the treatment of cognitive disabilities (e.g., Morrison, 2007). Holt, Johnson, Yaglom, and Brenner (2015) devised a programme, PAWSitive Visits, for a residential home in which domestic and exotic animals – they cite "Dogs, cats, miniature horses, alpacas, exotic mammals, birds, and reptiles.... rabbits, chinchillas, pot-bellied pigs, hedgehogs, and owls (p. 272) – are brought into the home each week for the residents to interact with". The programme provides education for residents and for students on placement, encourages social interaction and prompts residents to engage their cognitive facilities by recalling their own pets.

Ebener and Oh (2017) reviewed 33 studies published between 1979 and 2013 concerned with animal-assisted interventions in long-term care facilities. They considered the advantages and disadvantages of resident or visiting different species. Thus, for example, visiting cats are the animal of choice for encouraging socialisation and conversation; visiting dogs sooth agitated behaviour and encourage reminiscing and socialisation. On the other hand, resident birds increase life satisfaction and reduce depression while fish assist in gaining weight. Ebener and Oh also consider the practicalities and costs of animals in care facilities. It is cheapest to have a visiting animal; fish and birds are the least costly resident choice. It is even cheaper to play recordings of fish or to have a robotic animal. The main issue with cats and dogs is the risk of allergies and aggression, some residents' fear of animals, and the possibility that the animals may become stressed in this environment.

Criminals

Strimple (2003) makes the point that animals have a long history as part of prison life. Some prisons have farms where prisoners work, including taking responsibility for animals, during their sentence. In the same way that education is seen as desirable for prisoners as it provides knowledge and skills that the criminal may use when released, so working with animals was perceived as giving prisoners an opportunity for positive change. Strimple describes how a wave of animal care initiatives, mainly with horses and dogs, appeared in American prisons. In recent times the emphasis has shifted from animal care to a more interventionist approach: there is greater use of animal programmes aimed to change the prisoner and so lower recidivism. In a review of Prison Animal Programmes (PAPs) in Australia, Mulcahy and McLaughlin (2013) raise the pertinent issue of evaluation. They observe that while PAPS have an intuitive appeal (which is true of many other non-evidence based interventions) the standard of their evaluation, where it exists, is far from optimum.

The use of AAI in prisons, particularly with dogs but also with horses (Bachi, 2013), has become increasingly popular in several countries including Australia, Canada, Italy, New Zealand, and the USA (Britton & Button, 2005; Humby & Barclay, 2019; Jalongo, 2019). Fournier, Geller, and Fortney (2007) describe the *PenPals* programme, an American dog-training scheme where dogs from local shelters are trained over an 8 to 10-week period by volunteer prisoners. The dogs live with the prisoners who are given dog-training skills, look after the dogs' food, shelter and grooming, and train them in basic obedience. Once the training is complete the dog is homed and the process restarts with a new shelter dog. Fournier, Geller, and Fortney used a pre-post design to compare the social skills and instructional infringements during the research period for the prisoner-trainers and a matched control group. The prisoner-trainers showed an improvement in social skills and fewer institutional infractions. Flynn, Combs, Gandenberger, Tedeschi, and Morris (2020) reported positive psychological changes, alongside fewer institutional infractions, in prisoners taking part in a dog training programme. These types of programme are popular with prisoners (Smith, 2019) although less so with prison staff who can show some resistance to their implementation despite their positive effect on institutional safety (Humby & Barclay, 2019).

The *PenPals* programme illustrates the range of potential outcome measures for prison-based interventions: there is the effect on the prisoner's psychological functioning and behaviour within the jail and the longer-term impact on behaviour in the community after release. There are two meta-analyses which have looked at the effects of dog training programmes in prisons which both broadly favour their effects on psychological functioning and criminal behaviour (Cooke & Farrington, 2016; Duindam et al. 2020).

While male adults form the large majority of the prison population, women and young offenders constitute a sizable minority. For women offenders, including those with mental health problems (Jasperson, 2010), there are indications that dog programmes may have beneficial psychological and behavioural outcomes (Cooke & Farrington, 2015; Montes et al. 2017).

Animals are a part of the life of many adolescents and may mitigate some of the risk factors for delinquency by assisting the young person to learn to take responsibility for another's welfare (see Chapter 2). Pelyva, Kresák, Szovák, and Tóth (2020) make the case that contact with animals, specifically horses in their case, reinforces prosocial behaviour in adolescents. School truancy and dropout is an established risk factor for delinquency and, while academic learning and qualifications may be desirable, keeping young people in school acts towards the prevention of time spent on the streets. Ho, Zhou, Fung, and Kua (2017) used equine-assisted learning for Singaporean youths at-risk for school failure. While the programme had positive effects on variables such as individual character traits, there were no measures of social behaviour.

Some adolescents will commit crimes and become young offenders who face the penalties imposed by the criminal justice system. In the juvenile justice system, the majority of animal assisted programmes with young offenders use dogs (e. g., Smith & Smith, 2019) although there are examples where horses are used (e.g., Hemingway, Meek, & Hill, 2015). The evidence for the effectives of animal assisted programmes with young offenders is not compelling. Two studies which used a strong, randomised, research design failed to find a significant effectiveness of the programmes on psychological functioning (Grommon, Carson, & Kenney, 2020; Shen, Xiong, Chou, & Hall, 2018).

It is evident that there is a growing and widespread use of AAT both in familiar contexts, such as treatment of mental health problems, and in extreme situations such as the welfare of refugees (Every et al., 2015) and preparation for natural disasters (Thompson et al., 2014). It is also clear from the literature that practitioners are enthusiastic about incorporating AAT into their practice. However, enthusiasm is not a measure of effectiveness: the pressing question is does AAT work?

AAT: Does it work?

An intervention may be called evidence-based if it follows one of two routes: (i) a theory is supported by robust evidence allowing a treatment to develop; (ii) the accumulated evidence base allows confidence in a particular way of working. In the case of AAT, we may ask is it supported by theory or evidence?

Theoretical rationale for ATT

Kruger and Serpell (2010) note that with respect to theory:

> A considerable variety of possible mechanisms of action have been proposed or alluded to in the literature, most of which focus on the supposedly unique intrinsic attributes of animals that appear to contribute to therapy. Others emphasize the value of animals as living instruments that can be used to affect positive changes in patients' self-concept and behavior through the acquisition of various skills, and the acceptance of personal agency and responsibility (p. 37).

Thus, Kruger and Serpell note that it may be that animals bring intrinsic qualities and behaviours which smooth the therapeutic path. In addition, animals can act as catalysts or mediators for social interaction between people thereby reinforcing the relationship between therapist and patient.

In the context of AAT for pain relief, Marcus (2013) notes that a therapy dog can precipitate physiological changes, such as lowering blood pressure and reducing stress hormones, which can lower pain. Busch et al. (2016) favour a psychosocial explanation for the effects of AAT on attention-deficit disorder. This account includes psychological factors such as the animal's effects on motivation and calming (which also has a physiological component), alongside cognitive effects such as improved attention. Shen, Xiong, Chou, and Hall (2018) carried out a systematic review of seven qualitative studies of AAI. They noted six themes, as shown in Table 7.1, which were "Related to possible mechanisms of intervention effectiveness" (p. 205).

Table 7.1 Possible mechanisms of AAI effectiveness (after Shen et al., 2018)

1. Fostering feelings of normality.
2. Improving activity levels.
3. Self-esteem enhancement.
4. Physical contact, belonging, and companionship.
5. Calming and comforting.
6. Distraction.

As Shen et al. point out, in keeping with Kruger and Serpell, the commonality across the six themes is contact with the animal. With particular reference to cardiovascular disease and depression Wells (2019) suggests that the mechanisms by which animals enhance human health lie in improving physical fitness, offering companionship, and facilitating social interaction.

Identifying the strengths and weakness of the various theories relies on robust evidence. However, as Wells points out, methodological variations across the AAT literature pose problems in trying to build a coherent picture.

The evidence base for ATT

When considering the evidence base for ATT, as with any intervention, there are key factors to consider such as target population, type of animal, methodology including choice of dependent and independent variables, and experimental design. This need for research rigour runs up against the pervasive view, evident in some of the AAT literature, that animals are good for you. We know that some people form strong attachments to their companion animal and, as Herzog (2011) notes, it is a short step from attachment to believing that pets have the power to heal. This belief appears to extend into the evaluative literature as seen, for example, in an evaluation by Holt et al. (2015) of the PAWSitive Visits programme: they state that this programme "Is a unique and successful program, coordinated in a manner that ensures the health and safety of both animal and human counterparts, while providing residents with experiences that enhance their well-being" (pp. 267–268); they then state that "Findings from this very small pilot study were not statistically significant" (p. 276). The first statement perhaps represents their preferred outcome; the second statement is what the science showed and flatly contradicts the first.

In an emerging field it is not unexpected that practitioners are enthusiastic about the effects of ATT and are keen to see its benefits. However, as Herzog states: "Personal convictions, however, do not constitute scientific evidence. Claims about the medical and psychological benefits of living with animals need to be subjected to the same standards of evidence as a new drug, medical device, or form of psychotherapy" (pp. 236–237). Kazdin (2017) provides a set of methodological practices – including randomisation of participants, the use of multiple assessment and outcome measures, and follow-up – used in psychotherapy research and which could be used with AAT. There are several systematic reviews and meta-analyses which provide a more reliable estimate of the effectiveness of AAT.

Systematic reviews

Signal, Taylor, Botros, Prentice, and Lazarus (2013) reviewed 14 studies of therapies which involved horses for a wide range of populations. When taking into account the various methodological shortfalls in these studies, they concluded that the evidence gives "Credibility to the employment of equine-assisted techniques as an adjunct to traditional interventions for populations with health challenges" (p. 428). Anestis, Anestis, Zawilinski, Hopkins, and Lilienfeld (2014) reviewed 14 studies of equine-related treatments specifically for mental disorders. They reach the stark conclusion that: "There is negligible evidence that it offers benefits to individuals with mental disorders or other psychological difficulties" (p. 1129). O'Haire, Guérin, and Kirkham (2015) focussed on ten studies evaluating the effects of AAT on trauma. There were good outcomes in terms of reduced symptomatology for anxiety, depression, and PTSD. However, O'Haire, Guérin, and Kirkham commented on the low level of methodological rigour which limits confidence in the generalisability of this application of ATT.

Maujean, Pepping, and Kendall (2015) considered the effects of eight AAT evaluations employing Randomized Controlled Trials (RCTs) which are held to be the gold standard in research designs. They were optimistic that AAT could benefit to a wide range of populations but knowledge is needed to determine what specific types of AAT are beneficial for specific populations. Kamioka et al. (2014) similarly conducted a systematic review of 11 evaluations of AAT which employed a RCT. They reach the view that "AAT may be an effective treatment for mental and behavioral disorders such as depression, schizophrenia, and alcohol/drug addictions" (p. 387). Bert et al. (2016) looked at the benefits and risks of the therapeutic use of animals. They concluded that benefits such as anxiety and stress reduction outweighed costs such as allergies and infections favouring the use of ATT.

Santaniello et al. (2020) conducted an "umbrella review" summarising the methodological issues highlighted in 15 systematic reviews of AAT. They found variations in descriptions of therapeutic parameters such as number and length of sessions and the duration of treatment, while there was little the uniformity in the descriptions of the intervention. There was also disparity in the research methodology applied to study outcome.

Meta-analyses

Nimer and Lundahl (2007) included 49 studies in a meta-analysis of AAT. Overall, across the studies AAT gave moderate effect sizes in positive outcomes in autism-spectrum symptoms, behavioral problems, emotional well-being and medical difficulties. They conclude that AAT shows promise as an adjunct to traditional interventions but that more evidence is needed to establish the optimum use of AAT. Wilkie, Germain, and Theule (2016) conducted a meta-analysis of seven studies of the outcome of equine therapy. The reported a medium effect size leading them to conclude that: "Overall, this meta-analysis found that participation in an equine therapy program effectively increased overall level of functioning among adolescent at-risk youth" (p. 388). Finally, Germain, Wilkie, Milbourne, and Theule (2018) included eight studies in a meta-analysis of AAT for people experiencing the effects of trauma. They found a large effect size with the additional finding that the effect of the therapy increased according to the percentage of females in the treatment group.

The are two particular issues identified in the systematic reviews and meta-analyses which help account for the confusing picture they produce. First, poor reporting of therapeutic protocols makes it extremely difficult to compare across studies; second, there is a preponderance of studies using weaker, poorly controlled research designs with small sample sizes. A way to manage the poor reporting of therapeutic procedures is to introduce treatment manuals which detail the steps taken to complete the intervention satisfactorily. There is a long use of manuals in clinical (e.g., Wilson, 2007) and forensic psychology (Hollin & Palmer, 2006) which could inform progress in this area. The need for stronger methodologies does not necessarily mean the use of randomised control trials which can be both expensive and complex to carry out. There are other quasi-experimental methodologies which can produce robust data (Hollin, 2008; Kazdin,

2017) alongside techniques such as propensity score analysis which are beginning to be used in evaluations of AAT (Miles et al., 2017). Finally, while AAT may have positive benefits in the short-term it is essential to know how long these effects last and therefore more long-term follow-up studies are needed.

References

Albertini, M., Mazzola, S., Sincovich, M., & Pirrone, F. (2016). Canine scent detection of lung cancer: Preliminary results. *Journal of Veterinary Science & Research, 1,* 000119.

Aldridge, R. W., Story, A., Hwang, S. W., Nordentoft, M., Luchenski, S. A., Hartwell, G.,… Hayward, A. C. (2018). Morbidity and mortality in homeless individuals, prisoners, sex workers, and individuals with substance use disorders in high-income countries: A systematic review and meta-analysis. *The Lancet, 391*(10117), 241–250.

Alexandra, S., Dimitrios, V., Meropi, T., Alexandros, B., Elissavet, F., Sofia, D., & Avraam, P. (2019). Effects of equine assisted activities and therapies in children with autism spectrum disorder: A systematic review. *Open Science Journal of Education, 7,* 6–13.

American Psychiatric Association. (1994). *Diagnostic and statistical manual of mental disorders* (4th). Arlington, VA: American Psychiatric Association.

American Psychiatric Association. (2013). *Diagnostic and statistical manual of mental disorders* (5th ed). Arlington, VA: American Psychiatric Association.

Anderson, S., & Meints, K. (2016). Brief report: The effects of equine-assisted activities on the social functioning in children and adolescents with autism spectrum disorder. *Journal of Autism and Developmental Disorders, 46,* 3344–3352.

Anestis, M. D., Anestis, J. C., Zawilinski, L. L., Hopkins, T. A., & Lilienfeld, S. O. (2014). Equine-related treatments for mental disorders lack empirical support: A systematic review of empirical investigations. *Journal of Clinical Psychology, 70,* 1115–1132.

Arnon, S., Fisher, P. W., Pickover, A., Lowell, A., Turner, J. B., Hilburn, A.,… Hamilton, A. (2020). Equine-assisted therapy for veterans with PTSD: Manual Development and preliminary findings. *Military Medicine, 185,* e557–e564.

Bachi, K. (2013). Equine-facilitated prison-based programs within the context of prison-based animal programs: State of the science review. *Journal of Offender Rehabilitation, 52,* 46–74.

Berget, B., Ekeberg, Ø., & Braastad, B. O. (2008). Animal-assisted therapy with farm animals for persons with psychiatric disorders: Effects on self-efficacy, coping ability and quality of life, a randomized controlled trial. *Clinical Practice and Epidemiology in Mental Health, 4,* 9.

Bert, F., Gualano, M. R., Camussi, E., Pieve, G., Voglino, G., & Siliquini, R. (2016). Animal assisted intervention: A systematic review of benefits and risks. *European Journal of Integrative Medicine, 8,* 695–706.

Bishop, M., & Allen, C. A. (2007). Coping with epilepsy: Research and interventions. In E. Martz & H. Livneh (Eds.), *Coping with chronic illness and disability: Theoretical, empirical, and clinical aspects* (pp. 241–266). Boston, MA: Springer.

Boss, L., Branson, S., Hagan, H., & Krause-Parello, C. (2019). A systematic review of equine-assisted interventions in military veterans diagnosed with PTSD. *Journal of Veterans Studies, 5,* 23–33.

Braun, C., Stangler, T., Narveson, J., & Pettingell, S. (2009). Animal-assisted therapy as a pain relief intervention for children. *Complementary therapies in clinical practice, 15,* 105–109.

Britton, D. M. & Button, A. (2005). Prison pups: Assessing the effects of dog training programs in correctional facilities. *Journal of Family Social Work, 9,* 79–95.

Brown, C. A., Wang, Y., & Carr, E. C. J. (2018). Undercover dogs: Pet dogs in the sleep environment of patients with chronic pain. *Social Sciences, 7,* 157.

Brown, S. W., & Goldstein, L. H. (2011). Can seizure-alert dogs predict seizures? *Epilepsy Research, 97,* 236–242.

Brown, S. W., & Strong, V. (2001). The use of seizure-alert dogs. *Seizure, 10,* 39–41.

Busch, C., Tucha, L., Talarovicova, A., Fuermaier, A. B. M., Lewis-Evans, B., & Tucha, O. (2016). Animal-assisted interventions for children with Attention Deficit/Hyperactivity Disorder: A theoretical review and consideration of future research directions. *Psychological Reports, 118*, 292–331.

Buszewski, B., Rudnicka, J., Ligor, T., Walczak, M., Jezierski, T., & Amann, A. (2012). Analytical and unconventional methods of cancer detection using odor. *Trends in Analytical Chemistry, 38*, 1–12.

Carr, E. C. J., Norris, J. M., Hayden, K. A., Pater, R., & Wallace, J. E. (2020). A scoping review of the health and social benefits of dog ownership for people who have chronic pain. *Anthrozoös, 33*, 207–224.

Carr, E. C. J., Wallace, J. E., Onyewuchi, C., Hellyer, P. W., & Kogan, L. (2018). Exploring the meaning and experience of chronic pain with people who live with a dog: A qualitative study. *Anthrozoös, 31*, 551–565.

Carr, S., & Rockett, B. (2017). Fostering secure attachment: Experiences of animal companions in the foster home. *Attachment & Human Development, 19*, 259–277.

Catala, A., Cousillas, H., Hausberger, M., & Grandgeorge, M. (2018). Dog alerting and/or responding to epileptic seizures: A scoping review. *PLoS One, 13*, e0208280.

Catala, A., Grandgeorge, M., Schaff, J. L., Cousillas, H., Hausberger, M., & Cattet, J. (2019). Dogs demonstrate the existence of an epileptic seizure odour in humans. *Scientific Reports, 9*, 1–7.

Cherner, R. A., Farrell, S., Hwang, S. W., Aubry, T., Klodawsky, F., Hubley, A. M.,... To, M. J. (2018). An investigation of predictors of mental health in single men and women experiencing homelessness in three Canadian cities, *Journal of Social Distress and the Homeless, 27*, 25–33.

Cleary, M., Visentin, D., Thapa, D. K., West, S., Raeburn, T., & Kornhaber, R. (2020). The homeless and their animal companions: An integrative review. *Administration and Policy in Mental Health and Mental Health Services Research, 47*, 47–59.

Clements, H., Valentin, S., Jenkins, N., Rankin, J., Baker, J. S., Gee, N.,... Sloman, K. (2019). The effects of interacting with fish in aquariums on human health and well-being: A systematic review. *PLoS One, 14*, e0220524.

Cole, K. M., Gawlinski, A., Steers, N., & Kotlerman, J. (2007). Animal-assisted therapy in patients hospitalized with heart failure. *American Journal of Critical Care, 16*, 575–585.

Contalbrigo, L., De Santis, M., Toson, M., Montanaro, M., Farina, L., Costa, A., & Nava, F. A. (2017). The efficacy of dog assisted therapy in detained drug users: A pilot study in an Italian attenuated custody institute. *International journal of environmental research and public health, 14*, 683.

Cooke, B. J., & Farrington, D. P. (2015). The effects of dog-training programs: Experiences of incarcerated females. *Women & Criminal Justice, 25*, 201–214.

Cooke, B. J., & Farrington, D. P. (2016). The effectiveness of dog-training programs in prison: A systematic review and meta-analysis of the literature. *The Prison journal, 96*, 854–876.

Cornu, J.-N., Cancel-Tassin, G., Ondet, V., Girardet, C., & Cussenot, O. (2011). Olfactory detection of prostate cancer by dogs sniffing urine: A step forward in early diagnosis. *European Urology, 59*, 197–201.

Cronley, C., Strand, E. B., Patterson, D. A., & Gwaltney, S. (2009). Homeless people who are animal caretakers: A comparative study. *Psychological Reports, 105*, 481–499.

Crowley-Robinson, P., Fenwick, D. C., & Blackshaw, J. K. (1996). A long-term study of elderly people in nursing homes with visiting and resident dogs. *Applied Animal Behaviour Science, 47*, 137–148.

Cruz-Fierro, N., Vanegas-Farfano, M., & González-Ramírez, M. T. (2019). Dog-assisted therapy and dental anxiety: A pilot study. *Animals : An Open Access Journal from MDPI, 9*, 512.

Dietz, T. J., Davis, D., & Pennings, J. (2012). Evaluating animal-assisted therapy in group treatment for child sexual abuse. *Journal of Child Sexual Abuse, 21*, 665–683.

Donley, A. M., & Wright, J. D. (2012). Safer outside: A qualitative exploration of homeless people's resistance to homeless shelters. *Journal of Forensic Psychology Practice, 12*, 288–306.

Duindam, H. M., Asscher, J. J., Hoeve, M., Stams, G. J. J., & Creemers, H. E. (2020). Are we barking up the right tree? A meta-analysis on the effectiveness of prison-based dog programs. *Criminal Justice and Behavior, 47*, 749–767.

Ebener, J., & Oh, H. (2017). A review of animal-assisted interventions in long-term care facilities. *Activities, Adaptation & Aging, 41*, 107–128.

Engelman, S. R. (2013). Palliative care and use of animal-assisted therapy. *Omega, 67*, 63–67.

Every, D., Smith, K., Smith, B., Trigg, J., & Thompson, K. (2017). How can a donkey fly on the plane? The benefits and limits of animal therapy with refugees. *Clinical Psychologist, 21*, 44–53.

Ferrigno, S. (2015). Survey on the relationship between homeless people and the dog. *Dog Behavior, 2*, 18–24.

Fine, A. H. (Ed.) (2015). *Handbook on animal-assisted therapy: Foundations and guidelines for animal-assisted interventions* (4th ed.). San Diego, CA: Academic Press.

Fitzpatrick, K., Myrstol, B. A., & Miller, E. (2015). Does context matter? Examining the mental health among homeless people. *Community Mental Health Journal, 51*, 215–221.

Flynn, E., Combs, K. M., Gandenberger, J., Tedeschi, P., & Morris, K. N. (2020). Measuring the psychological impacts of prison-based dog training programs and in-prison outcomes for inmates. *The Prison journal, 100*, 224–239.

Fournier, Geller, & Fortney, (2007). Human-animal interaction in a prison setting: Impact on criminal behaviour, treatment progress, and social skills. *Behavior and Social Issues, 16*, 89–105.

Friedmann, E., Katcher, A. H., Lynch, J. J., & Thomas, S. A. (1980). Animal companions and one-year survival of patients after discharge from a coronary care unit. *Public Health Reports, 95*, 307–312.

Friedmann, E., Thomas, S. A., Son, H., Chapa, D., & McCune, S. (2013). Pet's presence and owner's blood pressures during the daily lives of pet owners with pre-to mild hypertension. *Anthrozoös, 26*, 535–550.

Gabriels, R. L., Pan, Z., Dechant, B., Agnew, J. A., Brim, N., & Mesibov, G. (2015). Randomized controlled trial of therapeutic horseback riding in children and adolescents with autism spectrum disorder. *Journal of the American Academy of Child & Adolescent Psychiatry, 54*, 541–549.

Gatti, F., Walderhaug, E., Kern-Godal, A., Lysell, J., & Arnevik, E. A. (2020). Complementary horse-assisted therapy for substance use disorders: A randomized controlled trial. *Addiction Science and Clinical Practice, 15*, 1–11.

Germain, S. M., Wilkie, K. D., Milbourne, V. M. K., & Theule, J. (2018). Animal-assisted psychotherapy and trauma: A meta-analysis. *Anthrozoös, 31*, 141–164.

Grandjean, D., Sarkis, R., Tourtiere, J.-P., Julien-Lecocq, C., Benard, A., Roger, V. .... Desquilbet, L. (2020). Detection dogs as a help in the detection of COVID-19: Can the dog alert on COVID-19 positive persons by sniffing axillary sweat samples? Proof-of-concept study. *bioRxiv*, posted 5 June.

Green, H. D. Jr, Tucker, J. S., Golinelli, D., & Wenzel, S. L. (2013). Social networks, time homeless, and social support: A study of men on Skid Row. *Network Science (Cambridge University Press), 1*, 305–320.

Griffioen, R., van der Steen, S., Cox, R., Verheggen, T., & Enders-Slegers, M.-J. (2019). Verbal interactional synchronization between therapist and children with autism spectrum disorder during dolphin assisted therapy: Five case studies. *Animals : An Open Access Journal from MDPI, 9*, 716.

Grommon, E., Carson, D. C., & Kenney, L. (2020). An experimental trial of a dog-training program in a juvenile detention center. *Journal of Experimental Criminology, 16*, 299–309.

Guerrero-Flores, H., Apresa-Garcia, T., Garay-Villar, Ó., Sánchez-Pérez, A., Flores-Villegas, D., Bandera-Calderon, A.,… Mata, O. (2017). A non-invasive tool for detecting cervical cancer odor by trained scent dogs. *BMC Cancer, 17*, 79.

Hall, S. S., Wright, H. F. Hames, A., P. A. W. S. Team, & Mills, D. S. (2016). The long-term benefits of dog ownership in families with children with autism. *Journal of Veterinary Behavior, 3*, 46–54.

Handlin, L., Nilsson, A., Lidfors, L., Petersson, M., & Uvnäs-Moberg, K. (2018). The effects of a therapy dog on the blood pressure and heart rate of older residents in a nursing home. *Anthrozoös, 31*, 567–576.

Harper, C. M., Dong, Y., Thornhill, T. S., Wright, J., Ready, J., Brick, G. W., & Dyer, G. (2015). Can therapy dogs improve pain and satisfaction after total joint arthroplasty? A randomized controlled trial. *Clinical Orthopaedics and Related Research, 473*, 372–379.

Heerde, J. A., Pallotta-Chiarolli, M. (2020). "I'd rather injure somebody else than get injured": An introduction to the study of exposure to physical violence among young people experiencing homelessness. *Journal of Youth Studies*, 23, 406–429.

Hemingway, A., Meek, R., & Hill, C. E. (2015). An exploration of an equine-facilitated learning intervention with young offenders. *Society & Animals*, 23, 544–568.

Herzog, H. (2011). The impact of pets on human health and psychological well-being: Fact, fiction, or hypothesis? *Current Directions in Psychological Science*, 20, 236–239.

Ho, N. F., Zhou, J., Fung, D. S. S., & Kua, P. H. J. (2017). Equine-assisted learning in youths at-risk for school or social failure. *Cogent Education*, 4, 1334430.

Hoagwood, K. E., Acri, M., Morrissey, M., & Peth-Pierce, R. (2017). Animal-assisted therapies for youth with or at risk for mental health problems: A systematic review. *Applied developmental science*, 21, 1–13.

Hoffman, L., & Coffey, B. (2008). Dignity and indignation: How people experiencing homelessness view services and providers. *Social Science Journal*, 45, 207–222.

Hollin, C. R. (2008). Evaluating offending behaviour programmes: Does only randomisation glister? *Criminology & Criminal Justice*, 8, 89–106.

Hollin, C. R., & Palmer, E. J. (Eds). (2006). *Offending behaviour programmes: Development, application, and controversies*. Chichester, West Sussex: John Wiley & Sons.

Holt, S., Johnson, R. A., Yaglom, H. D., & Brenner, C. (2015). Animal assisted activity with older adult retirement facility residents: The PAWSitive visits program. *Activities, Adaptation & Aging*, 39, 267–279.

Howe, L., & Easterbrook, M. J. (2018). The perceived costs and benefits of pet ownership for homeless people in the UK: Practical costs, psychological benefits and vulnerability. *Journal of poverty*, 22, 486–499.

Hoy-Gerlach, J., Rauktis, M., & Newhill, C. (2020). (Non-human) animal companionship: A crucial support for people during the COVID-19 pandemic. *Society Register*, 4, 109–120.

Humby, L., & Barclay, E. (2019). Pawsitive solutions: An overview of prison dog programs in Australia. *The Prison journal*, 98, 580–603.

Hwang, S. W., Kirst, M. J., Chiu, S., Tolomiczenko, G., Kiss, A., Cowan, L., & Levinson, W. (2009). Multidimensional social support and the health of homeless individuals. *Journal of Urban Health*, 86, 791–803.

Irvine, L. (2013). Animals as lifechangers and lifesavers: Pets in the redemption narratives of homeless people. *Journal of Contemporary Ethnography*, 42, 3–30.

Irvine, L., Kahl, K. N., & Smith, J. M. (2012). Confrontations and donations: Encounters between homeless pet owners and the public. *The Sociological Quarterly*, 53, 25–43.

Jalongo, M. R. (Ed.). (2019). *Prison dog programs: Renewal and rehabilitation in correctional facilities*. Cham, Switzerland: Springer.

Jang, B., Song, J., Kim, J., Kim, S., Lee, J., Shin, H. Y.,... Joung, Y. S. (2015). Equine-assisted activities and therapy for treating children with attention-deficit/hyperactivity disorder. *Journal of Alternative and Complementary Medicine*, 21, 546–553.

Jasperson, R. A. (2010). Animal-Assisted Therapy with female inmates with mental illness: A case example from a pilot program. *Journal of Offender Rehabilitation*, 49, 417–433.

Johnson, R. A., Albright, D. L., Marzolf, J. R., Bibbo, J. L., Yaglom, H. D., Crowder, S. M.,... & Osterlind, S. (2018). Effects of therapeutic horseback riding on post-traumatic stress disorder in military veterans. *Military Medical Research*, 5, 3.

Jones, M. G., Rice, S. M., & Cotton, S. M. (2019). Incorporating animal-assisted therapy in mental health treatments for adolescents: A systematic review of canine assisted psychotherapy. *PLoS One*, 14, e0210761.

Kamioka, H., Okada, S., Tsutani, K., Park, H., Okuizumi, H., Handa, S.,... Honda, T. (2014). Effectiveness of animal-assisted therapy: A systematic review of randomized controlled trials. *Complementary Therapies in Medicine*, 22, 371–390.

Kasstan, B., Hampshire, K., Guest, C., Logan, J. G., Pinder, M., Williams, K.,... Lindsay, S. W. (2019). Sniff and tell: The feasibility of using bio-detection dogs as a mobile diagnostic intervention for asymptomatic malaria in sub-Saharan Africa. *Journal of Biosocial Science*, 51, 436–443.

Kazdin, A. E. (2017). Strategies to improve the evidence base of animal-assisted interventions. *Applied developmental science, 21,* 150–164.

Kemp, K., Signal, T., Botros, H., Taylor, N., & Prentice, K. (2014). Equine facilitated therapy with children and adolescents who have been sexually abused: A program evaluation study. *Journal of child and family studies, 23,* 558–566.

Kerman, N., Gran-Ruaz, S., & Lem, M. (2019). Pet ownership and homelessness: A scoping review. *Journal of Social Distress and the Homeless, 28,* 106–114.

Kern-Godal, A., Arnevik, E. O., Walderhaug, E., & Ravndal, E. (2015). Substance use disorder treatment retention and completion: A prospective study of horse-assisted therapy (HAT) for young adults. *Addiction Science and Clinical Practice, 10,* 1–12.

Kern-Godal, A., Brenna, I. H., Kogstad, N., Arnevik, E. A., & Ravndal, E. (2016a). Contribution of the patient–horse relationship to substance use disorder treatment: Patients' experiences. *International Journal of Qualitative Studies on Health and Well-Being, 11,* 31636.

Kern-Godal, A., Brenna, I. H., Arnevik, E. A., & Ravndal, E. (2016b). More than just a break from treatment: How substance use disorder patients experience the stable environment in horse-assisted therapy. *Substance Abuse: Research and Treatment, 10,* 99–108.

Kidd, S. A., Gaetz, S., & O'Grady, B. (2017). The 2015 National Canadian Homeless Youth Survey: Mental health and addiction findings. *Canadian Journal of Psychiatry, 62,* 493–500.

Kim, C. H. Y. (2019). Homelessness and animal companionship. In L. Kogan & C. Blazina (Eds.), *Clinician's guide to treating companion animal issues* (pp. 365–378). Cambridge, MA: Academic Press.

Kirton, A., Wirrell, E., Zhang, J., & Hamiwka, L. (2004). Seizure-alerting and -response behaviors in dogs living with epileptic children. *Neurology, 62,* 2303–2305.

Klemetsen, M. G., & Lindstrom, T. C. (2017). Animal-assisted therapy in the treatment of substance use disorders: A systematic mixed methods review. *Human-Animal Interaction Bulletin, 5,* 90–117.

Koskinen, A., Bachour, A., Vaarno, J., Koskinen, H., Rantanen S., Bäck, L., & Klockars, T. (2019). A detection dog for obstructive sleep apnea. *Sleep and Breathing, 23,* 281–285.

Koukourikos, K., Georgopoulou, A., Kourkouta, L., & Tsaloglidou, A. (2019). Benefits of animal assisted therapy in mental health. *International Journal of Caring Sciences, 12,* 1898–1905.

Kramer, C. K., Mehmood, S., & Suen, R. S. (2019). Dog ownership and survival: A systematic review and meta-analysis. *Circulation: Cardiovascular Quality and Outcomes, 12,* e005554.

Krause-Parello, C. A., & Kolassa, J. (2016). Pet therapy: Enhancing social and cardiovascular wellness in community dwelling older adults. *Journal of Community Health Nursing, 33,* 1–10.

Krause-Parello, C. A., Thames, M., Ray, C. M., & Kolassa, J. (2018). Examining the effects of a service-trained facility dog on stress in children undergoing forensic interview for allegations of child sexual abuse. *Journal of Child Sexual Abuse, 27,* 305–320.

Krittanawong, C., Kumar, A., Wang, Z., Jneid, H., Virani, S. S., & Levine, G. N. (2020). Pet ownership and cardiovascular health in the US general population. *American Journal of Cardiology, 125,* 1158–1161.

Kruger, K. A., & Serpell, J. A. (2006). Animal-assisted interventions in mental health: Definitions and theoretical foundations. In A. Fine (Ed.), *Handbook on animal-assisted therapy: Theoretical foundations and guidelines for practice* (2nd ed.) (pp. 33–48). San Diego, CA: Elsevier.

Kruger, K. A., & Serpell, J. A. (2010). Animal-assisted interventions in mental health: Definitions and theoretical foundations. In A. H. Fine (Ed.), *Handbook on animal-assisted therapy: Theoretical foundations and guidelines for practice* (3rd ed.) (p. 33–48). San Diego, CA: Academic Press.

Kwon, S., Sung, I. Y., Ko, E. J., & Kim, H. S. (2019). Effects of therapeutic horseback riding on cognition and language in children with autism spectrum disorder or intellectual disability: A preliminary study. *Annals of Rehabilitation Medicine, 43,* 279–288.

Labrecque, J., & Walsh, C. A. (2011). Homeless women's voices on incorporating companion animals into shelter services. *Anthrozoös, 24,* 79–95.

Lang, U. E., Jansen, J. B., Wertenauer, F., Gallinat, J., & Rapp, M. A. (2010). Reduced anxiety during dog assisted interviews in acute schizophrenic patients. *European Journal of Integrative Medicine, 2,* 123–127.

Lanning, B. A., & Krenek N. (2013). Examining effects of equine assisted activities to help combat veterans improve quality of life. *Journal of Rehabilitative Research and Development, 50*, xv–xxii.

Lem, M., Coe, J. B., Haley, D. B., Stone, E., & O'Grady, W. (2016). The protective association between pet ownership and depression among street-involved youth: A cross-sectional study. *Anthrozoös, 29*, 123–136.

Levine, G. N., Allen, K., Braun, L. T., Christian, H. E., Friedmann, E., Taubert, K. A.,... Richard A. Lange. (2013). Pet ownership and cardiovascular risk: A scientific statement from the American Heart Association. *Circulation, 127*, 2353–2363.

Machová, K., Poběrežský, D., Svobodová, I., & Vařeková, J. (2017). A dog's effect on clients' heart rate and blood pressure and the possibilities of its use in relaxation. *Journal of Nursing, Social Studies, Public Health and Rehabilitation, 3*, 146–152.

Machová, K., Procházková, R., Říha, M., & Svobodová, I. (2019). The effect of animal-assisted therapy on the state of patients' health after a stroke: A pilot study. *International journal of environmental research and public health, 16*, Article Number 3272.

Maharaj, N. (2016). Companion animals and vulnerable youth: Promoting engagement between youth and professional service providers. *Journal of Loss and Trauma, 21*, 335–343.

Marcus, D. A. (2013). The science behind animal-assisted therapy. *Current Pain and Headache Reports, 17*, article number 322.

Marcus, D. A. Bernstein, C. D., Constantin, J. M., Kunkel, F. A., Breuer, P., & Hanlon, R. B. (2012). Animal-assisted therapy at an outpatient pain management clinic. *Pain Medicine, 13*, 45–57.

Marcus, D. A., Bernstein, C. D., Constantin, J. M., Kunkel, F. A., Breuer, P., & Hanlon, R. B. (2013). Impact of animal-assisted therapy for outpatients with fibromyalgia. *Pain Medicine, 14*, 43–51.

Maugeri, A., Medina-Inojosa, J. R., Kunzova, S., Barchitta, M., Agodi, A., Vinciguerra, M., & Lopez-Jimenez, F. (2019). Dog ownership and cardiovascular health: Results from the Kardiovize 2030 Project. *Mayo Clinic Proceedings: Innovations, Quality & Outcomes, 3*, 268–275.

Maujean, A., Pepping, C. A., & Kendall, E. (2015). A systematic review of randomized controlled trials of animal-assisted therapy on psychosocial outcomes. *Anthrozoös, 28*, 23–36.

Mazzola, S. M., Pirrone, F., Sedda, G., Gasparri, R., Romano, R., Spaggiari, L., & Mariangela, A. (2020). Two-step investigation of lung cancer detection by sniffer dogs. *Journal of Breath Research, 14*, 026011.

McCullough, A., Ruehrdanz, A., & Jenkins, M. (2016). The use of dogs in hospital settings. Purdue University IN: *HABRI Central Briefs*.

McCulloch, M., Jezierski, T., Broffman, M., Hubbard, A., Turner, K., & Janecki, T. (2006). Diagnostic accuracy of canine scent detection in early- and late-stage lung and breast cancers. *Integrative cancer therapies, 5*, 30–39.

Miles, J. N. V., Parast, L., Babey, S. H., Griffin, B. A. & Saunders, J. M. (2017). A propensity-score-weighted population-based study of the health benefits of dogs and cats for children. *Anthrozoös, 30*, 429–440.

Mims, D., & Waddell, R. (2016). Animal assisted therapy and trauma survivors. *Journal of Evidence-Informed Social Work, 13*, 452–457.

Montes, A. G., Lopez-Rodo, L. M., Rodríguez, I. R., Dequigiovanni, G. S., Segarra, N. V., Sicart, R. M. M.,... García-Navarro, Á. A. (2017). Lung cancer diagnosis by trained dogs. *European Journal of Cardio-Thoracic Surgery, 52*, 1206–1210.

Morrison, M. (2007). Health benefits of animal-assisted interventions. *Complementary Health Practice Review, 12*, 51–62.

Mubanga, M., Byberg, L., Egenvall, A., Sundström, J., Magnusson, P. K. E., Ingelsson, E., & Fall, T. (2019). Dog ownership and cardiovascular risk factors: A nationwide prospective register-based cohort study. *BMJ Open, 9*:e023447.

Mulcahy, C., & McLaughlin, D. (2013). Is the tail wagging the dog? A review of the evidence for prison animal programs. *Australian Psychologist, 48*, 369–378.

Narang, J., Schwannauer, M., Quayle, E., & Chouliara, Z. (2019). Therapeutic interventions with child and adolescent survivors of sexual abuse: A critical narrative review. *Children and Youth Services Review, 107*, 104559.

Nilsson, S. F., Nordentoft, M., Fazel, S., & Laursen, T. M. (2020). Homelessness and police-recorded crime victimisation: A nationwide, register-based cohort study. *The Lancet Public Health, 5*, e333–e341.

Nimer, J., & Lundahl, B. (2007). Animal-Assisted Therapy: A meta-analysis. *Anthrozoös, 20*, 225–238.

O'Haire, M. E. (2013). Animal-assisted intervention for autism spectrum disorder: A systematic literature review. *Journal of Autism and Developmental Disorders, 43*, 1606–1622.

O'Haire, M. E. (2017). Research on animal-assisted intervention and autism spectrum disorder, 2012–2015. *Applied developmental science, 21*, 200–216.

O'Haire, M. E., Guérin, N. A., & Kirkham, A. C. (2015). Animal-assisted intervention for trauma: A systematic literature review. *Frontiers in Psychology, 6*, 1121.

O'Haire, M. E., McKenzie, S. J., Beck, A. M., & Slaughter, V. (2019). Social behaviors increase in children with autism in the presence of animals compared to toys. *PLoS One, 8*, e57010.

Pelyva, I. Z., Kresák, R., Szovák, E., & Tóth, Á. L. (2020). How equine-assisted activities affect the prosocial behavior of adolescents. *International journal of environmental research and public health, 17*, 2967.

Pickel, D., Manucy, G. P., Walker, D. B., Hall, S. B., Walker, J. C. (2004). Evidence for canine olfactory detection of melanoma. *Applied Animal Behaviour Science, 89*, 107–116.

Pruskowski, K. A., Gurney, J. M., & Cancio, L. (2020). Impact of the implementation of a therapy dog program on burn center patients and staff. *Burns, 46*, 293–297.

Reitzes, D. C., Crimmins, T. J., Yarbrough, J., & Parker, J. (2011). Social support and social network ties among the homeless in a downtown Atlanta park. *Journal of Community Psychology, 39*, 274–291.

Rew, L. (2000). Friends and pets as companions: Strategies for coping with loneliness among homeless youth. *Journal of Child and Adolescent Psychiatric Nursing, 13*, 125–140.

Rhoades, H., Winetrobe, H., & Rice, E. (2015). Pet ownership among homeless youth: Associations with mental health, service utilization and housing status. *Child Psychiatry and Human Development, 46*, 237–244.

Roncati, L. (2019). Inside a mystery of oncoscience: The cancer-sniffing pets. *Oncoscience, 6*, 376–377.

Santaniello, A., Dicé, F., Claudia Carratú, R., Amato, A., Fioretti, A., & Menna, L. F. (2020). Methodological and terminological issues in animal-assisted interventions: An umbrella review of systematic reviews. *Animals : An Open Access Journal from MDPI, 10*, 759.

Scanlon, L., McBride, A., & Stavisky, J. (in press). Prevalence of pet provision and reasons for including or excluding animals by homelessness accommodation services. *Journal of Social Distress and Homelessness*,

Schreiner, P. J. (2016). Emerging cardiovascular risk research: Impact of pets on cardiovascular risk prevention. *Current Cardiovascular Risk Reports, 10*, Article 8.

Shen, R. Z. Z., Xiong, P., Chou, U. I., & Hall, B. J. (2018). "We need them as much as they need us": A systematic review of the qualitative evidence for possible mechanisms of effectiveness of animal assisted intervention (AAI). *Complementary Therapies in Medicine, 41*, 203–207.

Signal, T., Taylor, N., Botros, H., Prentice, K., & Lazarus, K. (2013). Whispering to horses: Childhood sexual abuse, depression and the efficacy of equine facilitated therapy. *Sexual Abuse in Australia and New Zealand, 5*, 24–32.

Signal, T., Taylor, N., Prentice, K., McDade, M., & Burke, K. J. (2017). Going to the dogs: A quasi-experimental assessment of animal assisted therapy for children who have experienced abuse. *Applied developmental science, 21*, 81–93.

Smith, H. P. (2019). A rescue dog program in two maximum security prisons: A qualitative study. *Journal of Offender Rehabilitation, 58*, 305–326.

Smith, H. P., & Smith, H. (2019). A qualitative assessment of a dog program for youth offenders in an adult prison. *Public Health Nursing, 36*, 507–513.

Stefanini, M. C., Martino, A., Allori, P., Galeotti, F., & Tani, F. (2015). The use of Animal-Assisted Therapy in adolescents with acute mental disorders: A randomized controlled study. *Complementary therapies in clinical practice, 21*, 42–46.

Stefanini, M. C., Martino, A., Bacci, B., & Tani, F. (2016). The effect of animal-assisted therapy on emotional and behavioral symptoms in children and adolescents hospitalized for acute mental disorders. *European Journal of Integrative Medicine, 8*, 81–88.

Strimple, E. O. (2003). A history of prison inmate–animal interaction programs. *American Behavioral Scientist, 47*, 70–78.

Strong, V., Brown, S., Huytens, M., & Coyle, H. (2002). Effect of trained Seizure Alert Dogs® on frequency of tonic–clonic seizures. *Seizure, 11*, 402–405.

Tan, V. X.-L., & Simmonds, J. G. (2019). Equine-assisted interventions for psychosocial functioning in children and adolescents with autism spectrum disorder: A literature review. *Journal of Autism and Developmental Disorders, 6*, 325–337.

Taverna, G., Tidu, L., Grizzi, F., Torri, V., Mandressi, A., Sardella, P.,... Hurle, R. (2015). Olfactory system of highly trained dogs detects prostate cancer in urine samples. *The Journal of Urology, 193*, 1382–1387.

Taylor, H., Williams, P., & Gray, D. (2004). Homelessness and dog ownership: An investigation into animal empathy, attachment, crime, drug use, health and public opinion. *Anthrozoös, 17*, 353–368.

Taylor, K. M., & Sharpe, L. (2008). Trauma and post-traumatic stress disorder among homeless adults in Sydney. *Australian and New Zealand Journal of Psychiatry, 42*, 206–213.

Thompson, K., Every, D., Rainbird, S., Cornell, V., Smith, B., & Trigg, J. (2014). No pet or their person left behind: Increasing the disaster resilience of vulnerable groups through animal attachment, activities and networks. *Animals : An Open Access Journal from MDPI, 4*, 214–240.

Trzmiel, T., Purandare, B., Michalak, M., Zasadzka, E., & Pawlaczyk, M. (2019). Equine assisted activities and therapies in children with autism spectrum disorder: A systematic review and a meta-analysis. *Complementary Therapies in Medicine, 42*, 104–113.

van Dartel, D., Schelhaas, H. J., Colon, A. J., Kho, K. H., & de Vos, C. C. (2020). Breath analysis in detecting epilepsy. *Journal of Breath Research, 14*, 031001.

van Houtert, E. A. E., Endenburg, N., Wijnker, J. J., Rodenburg, B., & Vermetten, E. (2018). The study of service dogs for veterans with Post-Traumatic Stress Disorder: A scoping literature review. *European Journal of Psychotraumatology, 9*(sup3), 1503523.

Waite, T. C., Hamilton, L., & O'Brien, W. (2018). A meta-analysis of animal assisted interventions targeting pain, anxiety and distress in medical settings. *Complementary therapies in clinical practice, 33*, 49–55.

Ward, S. C., Whalon, K., Rusnak, K., Wendell, K., & Paschall, N. (2013). The association between therapeutic horseback riding and the social communication and sensory reactions of children with autism. *Journal of Autism and Developmental Disorders, 43*, 2190–2198.

Welfare, H. R., & Hollin, C. R. (2015). Childhood and offence-related trauma in young people imprisoned in England and Wales for murder and other acts of serious violence: A descriptive study. *Journal of Aggression, Maltreatment & Trauma, 24*, 955–969.

Wells, D. L. (2019). The state of research on human–animal relations: Implications for human health. *Anthrozoös, 32*, 169–181.

Wenzel, S. L., Koegel, P., & Gelberg, L. (2000). Antecedents of physical and sexual victimization among homeless women: A comparison to homeless men. *American Journal of Community Psychology, 28*, 367–390.

White, J. H., Quinn, M., Garland, S., Dirkse, D., Wiebe, P., Hermann, M., & Carlson, L. E. (2015). Animal-assisted therapy and counseling support for women with breast cancer: An exploration of patient's perceptions. *Integrative cancer therapies, 14*, 460–467.

Wijker, C., Leontjevas, R., Spek, A., & Enders-Slegers, M.-J. (2020). Effects of dog assisted therapy for adults with autism spectrum disorder: An exploratory randomized controlled trial. *Journal of Autism and Developmental Disorders, 50*, 2153–2163.

Wilkie, K. D., Germain, S., & Theule, J. (2016). Evaluating the efficacy of equine therapy among at-risk youth: A meta-analysis. *Anthrozoös, 29*, 377–393.

Willis, C. M., Church, S. M., Guest, C. M., Cook, W. A., McCarthy, N., Bransbury, A. J.,... Church, J. C. T. (2004). Olfactory detection of human bladder cancer by dogs: Proof of principle study. *British Medical Journal, 329*(7468), 712.

Wilson, G. T. (2007). Manual-based treatment: Evolution and evaluation. In T. A. Treat, R. R. Bootzin, & T. B. Baker (Eds.), *Psychological clinical science: Papers in honor of Richard M. McFall* (pp. 105–132). New York: Psychology Press.

Wright, H., Hall, S., Hames, A., Hardiman, J., Mills, R., PAWS Project Team, & Mills, D. (2015a). Pet dogs improve family functioning and reduce anxiety in children with autism spectrum disorder. *Anthrozoös, 28*, 611–624.

Wright, H. F., Hall, S., Hames, A., Hardiman, J., Mills, R., PAWS Project Team, & Mills, D. S. (2015b). Acquiring a pet dog significantly reduces stress of primary carers for children with autism spectrum disorder: A prospective case control study. *Journal of Autism and Developmental Disorders, 45*, 2531–2540.

Xu, S.-L., Trevathan, E., Qian, Z., Vivian, E., Yanga, B.-Y., Li-Wen Hua,… Dong, G.-H. (2017). Prenatal and postnatal exposure to pet ownership, blood pressure, and hypertension in children: The Seven Northeastern Cities study. *Journal of Hypertension, 35*, 259–265.

Yamamoto, M. & Hart, L. A. (2019). Providing guidance on psychiatric service dogs and emotional support animals. In L. Kogan and C. Blazina (Eds.), *Clinician's guide to treating companion animal issues: Addressing human-animal interaction* (pp. 77–101). San Diego, CA: Elsevier.

Yeh, T.-L., Lei, W.-T., Liu, S.-J., & Chien, K.-L. (2019). A modest protective association between pet ownership and cardiovascular diseases: A systematic review and meta-analysis. *PLoS One, 14*, e0216231.

8 Eating, hurting, and killing animals

Animals bring many positives to human lives, from companionship and entertainment to preserving our safety and helping us recover from health problems. However, there is a dark side to the human–animal relationship: this chapter considers four ways in which humans are not the animal's best friend: (1) eating animals, (2) cruelty to animals, (3) animals for sport, and (4) animals in the laboratory.

Eating animals

As a self-professed nation of animal lovers, how do we bring ourselves to eat meat, especially given what we know about its production? There are several ways to allay the cognitive dissonance generated by the paradox of loving yet eating animals: we may see ourselves as dominant over animals, allowing us to control their fate, or deny that animals have any mental capacity so their slaughter is acceptable (Loughnan, Bastian, & Haslam, 2014). We are selective when it comes to which animals we eat: in the title to his 2010 book, Herzog considers why that is with regard to animals. *Some We Love, Some We Hate, Some We Eat* prompts two related issues: (1) Why do we eat some animals but not others? (2) Why do some people choose not eat meat or, indeed, eschew all animal products?

There are cultural and religious variations in which animals are eaten (e.g., Cohen, 2021; Frayne, 2018; Fuseini, Knowles, Hadley, & Wotton, 2017) and these social forces determine what is seen as acceptable. In some parts of the world, the idea of eating insects is viewed with abhorrence (Looy, Dunkel, & Wood, 2014), although this may be changing (Van Huis, 2020), whereas for man's best friend:

> Although dog meat is most often consumed in the form of a stew or soup (*tang*), it is also commonly taken in liquid form, *gaesoju*. Here, after the dog is killed, it is put into a stainless steel pressure cooker and boiled for up to 6 hours. The resulting liquefied dog is then mixed with herbs and strained into containers."
> (Podberscek, 2009, p. 619)

Some people will be repulsed by Podberscek's description of food preparation in South Korea. However, if we substitute beef or lamb for dog, then for most people the emotive content vanishes. We all hold our customs dear: Podberscek notes that Western attempts to interfere with dog consumption are seen by many Koreans as an attack on their culture. Dhont and Hodson (2014) found that in a sample of 260 Dutch-speaking male and female adults, those with right-wing political views consumed more meat as they perceived this as support for the dominant ideologies and displaying resistance to cultural change.

Some people reject the idea of eating meat: *vegetarians* do not eat meat but may consume, say, eggs and milk; *vegans* have a plant-based diet and avoid all animal products such as leather and wool. The decision not to eat meat may be related to having pets in childhood (Heiss & Hormes, 2018) and is likely to reflect the individual's moral values and liberal political beliefs, motivations for their own health and the wider environment, and concern about animal welfare (Rosenfeld, 2018; Rosenfeld & Burrow, 2017). The dietary choice not to eat meat can provoke a range of reactions, from admiration to disapproval, from other people (Earle & Hodson, 2017; Ruby et al., 2016).

Cruelty to animals

There are many everyday examples of cruelty to animals such as helicoptering in dog training (see Chapter 5), unnecessary medical procedures such as ear and tail docking (Mills, Robbins, & von Keyserlingk, 2016), and too many animals are beaten, neglected, hoarded, starved, kept in atrocious conditions, and worked to death. Cruelty to animals may begin in the child's upbringing: as children imitate what they observe, observation of animal cruelty, including within the home, may help explain the child's cruel behaviour. Given that violent behaviour is seldom restricted to a single target, it is not surprising that young people who are cruel to animals are likely to have observed domestic violence (Baldry, 2003; Currie, 2006) and themselves to have experienced parental corporal punishment and physical maltreatment (Flynn, 1999b; McEwen, Moffitt, & Arseneault, 2014).

In practice, assessment can be informed by The Cruelty to Animals Inventory (Dadds et al., 2004), noting that animal cruelty may be indicative of other childhood problems (Dadds, Turner, & McAloon, 2002). Petersen and Farrington (2007) observe that cruelty to animals is often a childhood feature of violent offenders. Kellert and Felthous (1985) interviewed imprisoned criminals who reported multiple motivations for animal cruelty ranging from teaching the animal a lesson, for the amusement of themselves and others, and hurting "bad" animals such as rats.

Flynn (1999a) surveyed a sample of 267 American university students asking about animal abuse and attitudes and experiences of family violence. Flynn reported that over one-sixth of the sample said they had abused an animal, ranging from killing and inflicting pain to sexual activity. The students who had been cruel to animals held more favourable attitudes towards spanking children and were more likely to approve of a husband slapping his wife than their non-abusing peers. While many people abuse animals, it is less certain which of them progress to violence towards other people (Hollin, 2016).

Animals for sport

Activities such as horse, greyhound, and pigeon racing are widespread, typically providing an opening for betting. Human ingenuity in laying a wager knows no bounds, exploiting the most unlikely animals: The Humane Society of the United States (HSUS) pleaded for an end to armadillo racing in Dallas, Texas (Humane Society of the United States, Gulf States Regional Office, 1981). Activities such as horse racing are regulated to attempt to minimise the risk of harm to the animal, although the animals' fate when their sporting career is over is quite another matter. There are other so-called sports where harm to the animal plays an intrinsic role: first, there are sports where animals inflict harm on each other as, for example, in cock-fighting and dogfighting to provide people with entertainment and an opportunity to gamble; second, in events such as hunting, often referred to a *blood* or *field sports;* and in bullfighting the animals *are* the sport to be injured and killed by people for their entertainment.

Animal versus animal

We humans have the ingenuity to pit animal against animal – Malchrowicz-Mośko, Munsters, Korzeniewska-Nowakowska, and Gravelle (2020) note cockfighting and camel wrestling – however, dogfighting is the most widespread of these activities. Organised dogfighting is illegal in many countries, including France, Germany, Great Britain, Italy, Russia, and the USA, but is a legal and a popular form of entertainment in regions of Latin America, Africa, the Middle East, and Asia. In organised dogfights, two or more dogs, typically Bull Terriers, are set against one another in a ring or pit. Intarapanich et al. (2017) compared the medical records of the injuries dogs received in spontaneously occurring dogfights and in illegally organised dogfights. The nature of the injuries differed significantly with dogs from organised fights more often having multiple wounds, particularly around the throat and head. In countries where dogfighting is illegal, knowledge of patterns of injury and scarring (Miller et al., 2016) can assist clinicians in identifying dogs injured in organised dogfights and so help to prosecute this crime. In accordance with Intarapanich et al. and Miller et al., an American survey of 200 prosecuting attorneys found that they most often relied upon detailed medical and crime scene reports, alongside photographic evidence, when making decisions about whether or not to proceed with a case and in successful prosecutions (Lockwood, Touroo, Olin, & Dolan, 2019).

Human versus animal

Harmful interactions between people and animals can take various forms. Sometimes the interaction is a (legal) spectator sport, as with bullfighting; at other times, as with hunting, it can be primarily for the enjoyment of those involved. There are other idiosyncratic localised contests: *La Rapa das Bestas*, for example, is a popular festival of horse wrestling in Galicia, Spain. Wild horses are captured and taken to the village, where the sport is to cut the horses' manes which leads to fights between human and horse.

Spectator sport

Bullfighting is legal in three European countries, France, Portugal, and Spain, and five Latin American countries, Colombia, Ecuador, Mexico, Venezuela, and Peru. In these countries, the bull is killed after the fight, sometimes in front of the spectators. There are variations of bullfighting in other countries including India and the USA. A Spanish study by Graña et al. (2004) asked 240 children aged 8 to 12 years for their views on bullfighting. The majority of children thought bullfighting was a violent event and one-half said they had seen televised bullfight. It may be asked, if it's evident to children that bullfighting is a violent event, why do people wish to watch an animal suffer and die? The answers to these questions are revealed by María, Mazas, Zarza, and de la Lama (2017) in a survey of 1,256 male and 1,266 female Spanish citizens asking for their views on bullfighting. This survey captured information on six aspects of bullfighting that highlight the relevant issues: (1) an individual liking of bullfighting as reflected in attendance at bullfights and watching them on television; (2) concern with bullfighting as a symbol of national culture, art, and identity; (3) the socio-economic advantages of bullfighting both locally in terms of jobs and local entertainment and nationally in loss of tourism; (4) concern that without bullfighting the Lidia cattle breed would disappear, meaning the loss of the Lidia bull "a brave and noble animal born to die in the bullfighting arena"; (5) the bull's instinct means it does not suffer in the fight; and (6) cultural evolution so that as societies become more mature then there is greater concern for animal welfare and a case for banning bullfighting as in other countries.

If we consider the bull, it is highly doubtful that, even if it is specifically bred for fighting, that it does not suffer during bullfighting (Andrade, 2018; Rodrigues & Achino, 2017). The cultural argument, evident in counties other than Portugal and Spain, such as India (Jayashree, Aram, & Ibrahim, 2019), requires closer inspection. Giménez-Candela (2019) makes the point that:

> I doubt that cultural is immovable. Culture is life and, for this reason, invariably changes. In this sense, that which had significance and value yesterday can, with no dishonour, no longer have it today. This is the case for public shows with animals on the occasion of parties or celebrations. Nowadays, despite its proliferation, they are anachronistic; they do not respond to the sensitivity or the values of a country that rejects and punishes violence in all forms, minus one; violence against animals in public shows. (p. 12)

In fact, Giménez-Candela's argument against cruelty extends beyond bullfighting, taking in other forms of public maltreatment of bulls: "Bulls and bull calves, which are forced 'actors' in the festivals in which they are made to run, are tethered, are thrown into the sea, their horns covered with burning hot tar, are driven with spears through the countryside, are jumped on or used for measuring the agility of the locals that celebrate these fiestas" (p. 11).

Hunting

In the modern world, there are very few people who must hunt to survive, yet hunting persists in many countries. Some hunting (not always legal, such as killing elephants and whales) takes place for commercial reasons as with animals killed for their pelts. The animals commercially hunted for their fur include beaver, bobcat, coyote, lynx, mink, marten, muskrat, opossum, otter, raccoon, skunk, and weasel. Commercial hunting is prevalent in many countries, although animals are also bred in captivity for their coats. However, the majority of hunting is not for commercial gain but for recreation. The practice of *recreational hunting* typically involves killing smaller mammals, such as deer, foxes, hares, and rabbits, as well as birds including ducks, grouse, pheasants, and even songbirds. *Trophy hunting* is a variant of recreational hunting where the aim is to kill larger animals such as elephants, lions, and sharks and take their heads and skins.

Table 8.1 Examples of the most-hunted species in the UK in 2016 (from Aebischer, 2019)

Birds	Estimated bag size
Grouse	650,000
Mallard	790,000
Woodpigeon	1,900,000
Mammals	
Fox	89,000
Grey Squirrel	150,000
Rabbit	350,000

It is impossible to know hunting's global death toll, but local estimates can be made. Aebischer (2019) examined the figures for the UK:

> A large number of bird and mammal species are legally shot or trapped in the UK. These include typical quarry species such as many waterfowl (Anatidae), most gamebirds (Phasianidae, Tetraonidae), some waders (Scolopacidae, Charadriidae), some rails (Rallidae), deer (Cervidae), boar (Suidae) and lagomorphs (Leporidae). They also include species that are often viewed as pests, such as most corvids (Corvidae), some gulls (Laridae), the fox *Vulpes vulpes*, some small mustelids (Mustelidae), the brown rat *Rattus norvegicus* and the introduced grey squirrel *Sciurus carolinensis*. (p. 1)

Aebischer calculates the 2016 "bags" for a range of species: some examples of the numbers of animals killed are given in Table 8.1.

Fishing

When we think of hunting it is animals such as foxes, deer, tigers, and lions that first come to mind. Should fish be included in this list? We may perceive big-game fishing for say marlin and tuna as a form of hunting, but does trout fishing fall into the same category? The sport of angling, both game and coarse with their divergent perspectives (Mordue & Wilson, 2018), is highly popular both as a hobby and as a competitive event, with thousands of anglers spending a great deal of money on equipment, licences, tourist fishing, and so on (Mordue, 2016). However, the fact remains that an animal is taken from its natural environment and killed for the angler's enjoyment. It may be thought that "catch and release" fishing at least allows the animal to live but the fish may be physically damaged, depending on factors such as the handling of the fish and the type of hook used (Arlinghaus et al., 2007).

Hunting methods

The hunter uses a variety of methods including animals, such as dogs and birds of prey, and equipment such as guns, traps, bows and arrows, fishing rods, and traps. In the digital age, the *cyber hunter* (also known as *internet hunting* and *remote-control hunting*) does not have to move from their armchair. The cyber hunter logs onto a website that gives access to an online webcam showing penned animals and gives access to remotely controlled firearms. The hunter uses their mouse or joystick to aim and shoot at the captive animals.

Those in favour argue that cyber hunting enables disabled people to hunt; the opposition comes, as expected, from anti-hunting groups but also from hunters on the grounds that as there is no fair chase, then this activity is not hunting.

Why hunt?

Presser and Taylor (2011) asked an American hunter to list what he liked about hunting and six themes emerged: (i) the challenge, (ii) being present in the moment, (iii) excitement, (iv) learning, (v) beauty of the surroundings, and (vi) social bonding. A notable point about this list is the absence of killing, so that the activity is transformed into "serious leisure," with the hunt seemingly secondary to communing with nature. Indeed, commentators such as Franklin (2001) see hunters as connected with the landscape rather than passing through like tourists.

The defence of hunting

There are several lines of defence for recreational hunting. As with bullfighting, there is the cultural argument that hunting is a tradition which should be upheld. Thus, the fox hunting debate has been recast as a conflict between town and country, whereby townies have no comprehension of rural customs (see Hillyard, 2007) so their opposition to hunting is in fact opposition to an integral aspect of country life (Milbourne, 2003). Regardless of the accuracy of this position, the same argument as with bullfighting applies: customs and traditions are not cast in stone and can and do change with the times.

Public attitudes to hunting

Public attitudes to hunting vary within and between counties and may be influenced by factors such as whether the hunting is for recreation or for providing food (e.g., Ankeny & Bray, 2018; Gamborg & Jensen, 2017; Krokowska-Paluszak et al., 2020). The hunting of animals for trophies has a long history and an incident involving the death of a lion named Cecil shone a light on public option of this practice (Figure 8.1). Godoy (2020) describes what happened:

> In the summer of 2015, Walter J. Palmer, a dentist from Minnesota, made a trip to the Hwange District in northwestern Zimbabwe. He hired professional hunter Theo Bronkhorst for US$50,000 to help him hunt and kill a lion. ... They drove to a legal hunting ground on the edge of the Hwange National Park. Lion-hunting isn't allowed in the park; however, the area immediately surrounding the park is part of their natural range. ... Palmer and Bronkhorst used bait to attract a male lion, a common practice when hunting them, to Antoinette Farm. At approximately 10 pm, on July 1, Palmer shot the lion with a bow and arrow wounding him. About eleven hours later, they managed to track down the bleeding animal and kill him. (pp. 759–760)

It so happened that Cecil was part of a study by Oxford University's Wildlife Conservation Research Unit and wore a GPS tracking collar, which the hunters destroyed after collecting Cecil's head and skin. Palmer did not face charges for his crime despite a previous conviction for a similar offence in America involving a black bear.

Figure 8.1 Cecil.
Source: Daughter#3 via flickr.

Cecil's death attracted a deluge of comments on social media – the number of mentions on Facebook, Twitter, and YouTube was estimated at over 90,000 – and articles in the worldwide press peaked at over 11,000 on a single day. The Oxford research group considered this huge response to Cecil's death and suggests it expresses an interest in wildlife and conservation and a repulsion not just for trophy hunting but for killing animals for sport (Feber, Johnson, & Macdonald, 2020; Macdonald, Jacobsen, Burnham, Johnson, & Loveridge, 2016).

Why do people feel this need to kill animals for trophies? A need that extends even to endangered animals such as giant pandas (Montgomery et al., 2020), which is ironically the symbol of The World Wildlife Fund. There is undoubtably a mixture of social and psychological factors involved in trophy hunting (Beattie, 2020), including a sense of achievement, affiliation with other like-minded people, and personal appreciation of the experience (Ebeling-Schuld & Darimont, 2017).

While some countries have laws prohibiting various forms of hunting, it is evident that animals are killed for sport in large numbers. Those opposed to hunting fall into two philosophical camps, as Mkono (2019) states: "First, the Kantian camp opposes trophy hunting regardless of any positive outcomes. There is for them an absolute moral imperative not to hunt for recreation. The second camp are the consequentialists, or more specifically, the utilitarian view traceable to the philosophy of Jeremy Bentham, wherein the morality of an action is judged on its outcomes (as opposed to motive or nature of the action)" (p. 213).

The absolute nature of the Kantian position is perfectly plain: animals are not commodities to be abused for human pleasure. Thus, as Batavia et al. (2019) state, it is morally inappropriate for a person to pay a fee to kill an animal and take its body as a trophy. As Mkono (2019) notes, the discourse of hunters taking Bentham's utilitarian position employ several themes to justify their behaviour: these themes are based on *altruism* in that hunting has good outcomes as with *animal conservation* (although not for the animal that's killed) and assisting in *sustainability* and *population control*. The use of *euphemisms* to dilute the violence inherent in hunting is found in terms such *taking* or *harvesting* rather than *killing*. While *anti-emotionality* portrays critics of hunting as irrational, emotional and sentimental, unlike the level-headed, responsible hunter. These rationalisations can be used to justify law-breaking as seen in the strategies used to support the illegal culling of badgers. It is held that as badger culling is necessary for the greater good so laws to the contrary are plainly unnecessity and irrelevant: this argument allows strong rebukes against those who oppose culling while and appealing to community loyalties (Enticott, 2001).

Vivisection

There are millions of animals, including cats, dogs, fish, horses, mice, monkeys, and rats, experimented upon every day in laboratories around the world. The aims of animal research are wide and varied, from medical research to cosmetics testing, although all with the risk that the animals may be hurt or killed. There are libraries of writings given to the full spectrum of views from all animal experimentation is wrong to that given guidelines and safeguards, the ends justifies the means. In contemporary psychology, animal research, typically with rodents, is evident in animal modelling studies and in neuroscience. Laboratory research using animals such as chimpanzees and dogs has largely fallen away, partly because of ethical concerns partly due the financial costs of maintaining facilities. In Chapter 1, several landmark studies in psychology that used animal experimentation were discussed. How would these studies be judged against modern-day ethical standards for human–animal relationships?

Pavlov and Seligman

The behavioural aspects of Pavlov's studies are unremarkable, although the measurement of saliva, originally part of his physiological research, involved an intrusive procedure. Kopaladze (2000) notes that Pavlov was aware of the ethical issues in animal experimentation and took them into accord in his research. Although, as commentators have stressed (Adams, 2020a, 2020b), the dogs may well have had a more miserable tale to tell than is evident from the textbooks. Seligman's experiments in learned helplessness have drawn criticism for inflicting pain and distress on the dogs. However, as in medical research, the defence rests on the findings, prompting new conceptualisations and treatments for psychological problems.

Skinner

The use of rats in laboratory research remains commonplace in psychological research, but there has been a shift from running mazes and the Skinner box to more invasive procedures. There are arguments that rats find the laboratory highly emotionally disturbing and their welfare suffers accordingly (Makowska & Weary, 2013).

Kohler, Gardner, and Harlow

The work of Kohler and Gardner is arguably closer to ethology than experimental psychology, although the animals are removed from their natural environment. Harlow's research is more open to criticism on the grounds of cruelty to the infant monkeys and the anxiety suffered by the mothers separated from their offspring. Harlow's experiments are sometimes justified as providing a valuable insight into the development of attachment and social behavior. At the time of the research, there was a dominant belief that attachment was related to physical (i.e., food) rather than emotional care. A utilitarian stance would argue that the benefits of the research in influencing the further study of attachment, which ultimately led to changes in considerations of the emotional care of those in hospitals and residential and day care, outweigh the costs borne by the animals. The use of chimpanzees highlights the ethical dilemmas, common to all animal research, in seeking to justify invasive procedures by their scientific method or anticipated consequences (Carvalho, Gaspar, Knight, & Vicente, 2019; Russell, 1997).

References

Adams, M. (2020a). The kingdom of dogs: Understanding Pavlov's experiments as human–animal relationships. *Theory & Psychology, 30,* 121–141.

Adams, M. (2020b, June). The kingdom of dogs. *The Psychologist,* 76–79.

Aebischer, N. J. (2019). Fifty-year trends in UK hunting bags of birds and mammals, and calibrated estimation of national bag size, using GWCT's National Gamebag Census. *European Journal of Wildlife Research, 65,* 64.

Andrade, G. (2018). Francis Wolff's flawed philosophical defense of bullfighting. *Between the Species, 22,* 158–184.

Ankeny, R., & Bray, H. (2018). Ferals or food? Does hunting have a role in ethical food consumption in Australia? In N. Carr & J. Young (Eds.), *Wild animals and leisure: Rights and wellbeing* (pp. 210–224). Milton Park, Oxfordshire: Routledge.

Arlinghaus, R., Cooke, S. J., Lyman, J., Policansky, D., Schwab, A., Suski, C.,... Thorstad, E. B. (2007). Understanding the complexity of catch-and-release in recreational fishing: An integrative synthesis of global knowledge from historical, ethical, social, and biological perspectives. *Reviews in Fisheries Science, 15,* 75–167.

Baldry, A. C. (2003). Animal abuse and exposure to interparental violence in Italian youth. *Journal of Interpersonal Violence, 18,* 258–281.

Batavia, C., Nelson, M. P., Darimont, C. T., Paquet, P. C., Ripple, W. J., & Wallach, A. D. (2019). The elephant (head) in the room: A critical look at trophy hunting. *Conservation Letters,* 12:e12565.

Beattie, G. (2020). *Trophy hunting: A psychological perspective.* Milton Park, Oxfordshire: Routledge.

Carvalho, C., Gaspar, A., Knight, A., & Vicente, L. (2019). Ethical and scientific pitfalls concerning laboratory research with non-human primates, and possible solutions. *Animals: An Open Access Journal from MDPI, 9,* 12.

Cohen, A. B. (2021). You can learn a lot about religion from food. *Current Opinion in Psychology, 40,* 1–5.

Currie, C. L. (2006). Animal cruelty by children exposed to domestic violence. *Child Abuse & Neglect, 30,* 425–435.

Dadds, M. R., Turner, C. M., & McAloon, J. (2002). Developmental links between cruelty to animals and human violence. *Australian and New Zealand Journal of Criminology, 35,* 363–382.

Dadds, M. R., Whiting, C., Bunn, P., Fraser, J. A., Charlson, J. H., & Pinola-Merlo, A. (2004). Measurement of cruelty in children: The cruelty to animals inventory. *Journal of Abnormal Child Psychology, 32,* 321–334.

Dhont, K., & Hodson, G. (2014). Why do right-wing adherents engage in more animal exploitation and meat consumption? *Personality and Individual Differences, 64*, 12–17.

Earle, M., & Hodson, G. (2017). What's your beef with vegetarians? Predicting anti-vegetarian prejudice from pro-beef attitudes across cultures. *Personality and Individual Differences, 119*, 52–55.

Ebeling-Schuld, A. M., & Darimont, C. T. (2017). Online hunting forums identify achievement as prominent among multiple satisfactions. *Wildlife Society Bulletin, 41*, 523–529.

Enticott, G. (2001). Calculating nature: The case of badgers, bovine tuberculosis and cattle. *Journal of Rural Studies, 17*, 149–164.

Feber, R. E., Johnson, P. J., & Macdonald, D. W. (2020). Shooting pheasants for sport: What does the death of Cecil tell us? *People and Nature, 2*, 82–95.

Flynn, C. P. (1999a). Animal abuse in childhood and later support for interpersonal violence in families. *Society & Animals, 7*, 161–172.

Flynn, C. P. (1999b). Exploring the link between corporal punishment and children's cruelty to animals. *Journal of Marriage and the Family, 61*, 971–981.

Franklin, A. (2001). Neo-Darwinian leisures, the body and nature: Hunting and angling in modernity. *Body & Society, 7*, 57–76.

Frayne, C. T. (2018). Animals in Christian and Muslim thought: Creatures, creation and killing for food. In A. Linzey & C. Linzey (Eds.), *The Routledge handbook of religion and animal ethics* (pp. 201–225). Milton Park, Oxfordshire: Routledge.

Fuseini, A., Knowles, T. G., Hadley, P. J., & Wotton, S. B. (2017). Food and companion animal welfare: The Islamic perspective. *CAB Reviews, 12*(043), 1–6.

Gamborg, C., & Jensen, F. S. (2017). Attitudes towards recreational hunting: A quantitative survey of the general public in Denmark. *Journal of Outdoor Recreation and Tourism, 17*, 20–28.

Giménez-Candela, M. (2019). Culture and animal mistreatment. *Animal Law (Forum of Animal Law Studies), 10*, 7–14.

Godoy, E. S. (2020). Sympathy for Cecil: Gender, trophy hunting, and the western environmental imaginary. *Journal of Political Ecology, 27*, 759–774.

Graña, J. L., Cruzado, J. A., Andreu, J. M., Muñoz-Rivas, M. J., Peña, M. E., & Brain, P. F. (2004). Effects of viewing videos of bullfights on Spanish children. *Aggressive Behavior, 30*, 16–28.

Heiss, S., & Hormes, J. M. (2018). Ethical concerns regarding animal use mediate the relationship between variety of pets owned in childhood and vegetarianism in adulthood. *Appetite, 123*, 43–48.

Herzog, H. (2010). *Some we love, some we hate, some we eat: Why it's so hard to think straight about animals.* New York: Harper Collins.

Hillyard, S. (2007). *The sociology of rural life.* Oxford: Berg.

Hollin, C. R. (2016). *The psychology of interpersonal violence.* Chichester, West Sussex: Wiley Blackwell.

Humane Society of the United States: Gulf States Regional Office 11. (1981, Fall). Report. http://animalstudiesrepository.org/gulstarn/11

Intarapanich, N. P., Touroo, R. M., Rozanski, E. A., Reisman, R. W., Intarapanich, P. P., & McCobb, E. C. (2017). Characterization and comparison of injuries caused by spontaneous versus organized dogfighting. *Journal of the American Veterinary Medical Association, 251*, 1424–1431.

Jayashree, B., Aram, A., & Ibrahim, Y. (2019). The voices of culture, conservation and in the media event around bullfight 'Jallikattu' in Tamil Nadu, India. *Journal of Media and Communication Studies, 11*, 20–30.

Kellert, S. R., & Felthous, A. R. (1985). Childhood cruelty to animals among criminals and non-criminals. *Human Relations, 38*, 1113–1129.

Kopaladze, R. A. (2000). Ivan P. Pavlov's view on vivisection. *Integrative Physiological and Behavioral Science, 35*, 266–271.

Krokowska-Paluszak, M., Łukowski, A., Wierzbicka, A., Gruchała, A., Sagan, J., & Skorupski, M. (2020). Attitudes towards hunting in Polish society and the related impacts of hunting experience, socialisation and social networks. *European Journal of Wildlife Research, 66*, 1–8.

Lockwood, R., Touroo, R., Olin, J., & Dolan, E. (2019). The influence of evidence on animal cruelty prosecution and case outcomes: Results of a survey. *Journal of Forensic Sciences, 64*, 1687–1692.

Looy, H., Dunkel, F. V., & Wood, J. R. (2014). How then shall we eat? Insect-eating attitudes and sustainable foodways. *Agriculture and Human Values, 31*, 131–141.

Loughnan, S., Bastian, B., & Haslam, N. (2014). The psychology of eating animals. *Current Directions in Psychological Science, 23*, 104–108.

Macdonald, D. W., Jacobsen, K. S., Burnham, D., Johnson, P. J., & Loveridge, A. J. (2016). Cecil: A moment or a movement? Analysis of media coverage of the death of a lion, Panthera leo. *Animals, 6*, 26.

Makowska, I. J., & Weary, D. M. (2013). Assessing the emotions of laboratory rats. *Applied Animal Behaviour Science, 148*, 1–12.

Malchrowicz-Mośko, E., Munsters, W., Korzeniewska-Nowakowska, P., & Gravelle, F. (2020). Controversial animal tourism considered from a cultural perspective. *Tourism, 30*, 21–30.

María, G. A., Mazas, B., Zarza, F. J., & de la Lama, G. C. M. (2017). Animal welfare, national identity and social change: Attitudes and opinions of Spanish citizens towards bullfighting. *Journal of Agricultural and Environmental Ethics, 30*, 809–826.

McEwen, F. S., Moffitt, T. E., & Arseneault, A. (2014). Is childhood cruelty to animals a marker for physical maltreatment in a prospective cohort study of children? *Child Abuse & Neglect, 38*, 533–543.

Milbourne, P. (2003). The complexities of hunting in rural England and Wales. *Sociologia Ruralis, 43*, 289–308.

Miller, K. A., Touroo, R., Spain, C. V., Jones, K., Reid, P., & Lockwood, R. (2016). Relationship between scarring and dog aggression in pit bull-type dogs involved in organized dogfighting. *Animals: An Open Access Journal from MDPI, 6*, 72.

Mills, K. E., Robbins, J., & von Keyserlingk, M. A. G. (2016). Tail docking and ear cropping dogs: Public awareness and perceptions. *PLoS One, 11*, e0158131.

Mkono, M. (2019). The trophy hunting controversy: How hunters rationalise their pastime in social media. In M. Mkono (Ed.), *Positive tourism in Africa* (pp. 211–229). Milton Park, Oxfordshire: Routledge.

Montgomery, R. A., Carr, M., Booher, C. R., Pointer, A. M., Mitchell, B. M., Smith, N., ... & Kramer, D. B. (2020). Characteristics that make trophy hunting of giant pandas inconceivable. *Conservation Biology, 34*, 915–924.

Mordue, T. (2016). Game-angling tourism: Connecting people, places and natures. *International Journal of Tourism Research, 18*, 269–276.

Mordue, T., & Wilson, S. (2018). Angler and fish relations in the UK: Ethics, aesthetics and material semiotics. In N. Carr & J. Young (Eds.), *Wild animals and leisure: Rights and wellbeing* (pp. 165–180). Milton Park, Oxfordshire: Routledge.

Petersen, M. L., & Farrington, D. P. (2007). Cruelty to animals and violence to people. *Victims and Offenders, 2*, 21–43.

Podberscek, A. L. (2009). Good to pet and eat: The keeping and consuming of dogs and cats in South Korea. *Journal of Social Issues, 65*, 615–632.

Presser, L., & Taylor, W. V. (2011). An autoethnography of hunting. *Crime, Law, and Social Change, 55*, 483–494.

Rodrigues, L. C., & Achino, E. (2017). A case study on moral disengagement and rationalization in the context of Portuguese bullfighting. *Polish Sociological Review, 3*(199), 315–317.

Rosenfeld, D. L. (2018). The psychology of vegetarianism: Recent advances and future directions. *Appetite, 131*, 125–138.

Rosenfeld, D. L., & Burrow, A. L. (2017). Vegetarian on purpose: Understanding the motivations of plant-based dieters. *Appetite, 116*, 456–463.

Ruby, M. B., Alvarenga, M. S., Rozin, P., Kirby, T. A., Richer, E., & Rutsztein, G. (2016). Attitudes toward beef and vegetarians in Argentina, Brazil, France, and the USA. *Appetite, 96*, 546–554.

Russell, D. (1997). Animal experimentation in psychology and the question of scientific merit. *Ethics and the Environment, 2*, 43–52.

Van Huis, A. (2020). Insects as food and feed, a new emerging agricultural sector: A review. *Journal of Insects as Food and Feed, 6*, 27–44.

9 Into the Anthropocene

There is more than one way to understand the long history of our planet. One method, broadly familiar if the intricates are rather less appreciated (e.g., Lewis & Maslin, 2015), is to subdivide geological eras in smaller units of time such as epochs and periods. After the formation of Earth about 4.6 billion years ago, there is the *Palaeozoic era*, beginning with the *Cambrian period* about 541 million years ago, and ending with the *Permian period* 252 million years ago. Earth moved to the *Mesozoic era* and the start of *Triassic period* and finally, 66 million years ago, the beginning of the *Cenozoic era* and the *Paleogene period*, bringing us 10,000 years ago to the present epoch, the *Holocene*. It has been proposed that we have entered a new epoch, the *Anthropocene* (*anthropo* meaning "man," *cene* meaning "new"), in which humans have become a major ecological and geological force and where unspoiled nature is almost gone (Crutzen & Stoermer, 2000). The sweeping use of the notion of the Anthropocene has met with some scepticism (e.g., Foster, 2019), but it is can be argued, Anthropocene or not, that we live at a time of significant climate change, precipitating environmental changes such as rising sea levels and large-scale wildfires. These global changes, aided and abetted by human activity, change and destroy habitats (Scanes, 2018), heightening the risk of animal extinction on a mass scale (Toukhsati, 2018).

There are various initiatives to stall and combat global change. On a large scale, there are international efforts such as the Kyoto Protocol that aim to reduce atmospheric pollution to slow the rate of climate change. Conservation can be helped by engaging people (and their money) in environmental tourism, including safaris and whale watching; there are ambitious breeding projects that aim to reintroduce species such as eagles and wolves into the wild. Founded in 1993, The Great Ape Project (https://www.projetogap.org.br/en/) is an international organization of specialists, including anthropologists and climatologists, advocating a UN Declaration of the Rights of Great Apes. The aim of the project is to confer basic legal rights on bonobos, chimpanzees, gorillas, and orangutans. Indeed, there is an international movement towards animal rights to complement human rights (Peters, 2020) although, judging by the quote Peters presents, there is a long way to go:

> In May 2018, US President Donald Trump spoke about illegal border crossings: 'We have people coming into the country, or trying to come in—and we're stopping a lot of them — but we're taking people out of the country. You wouldn't believe how bad these people are. These aren't people. These are animals.' Such dehumanisation (in this

case: of foreigners at the Californian-Mexican border) has — throughout history — been a standard discursive strategy to prepare, instigate, facilitate, and exculpate violence committed by humans against other humans. (p. 109)

On a smaller scale, each individual can take steps to reduce their impact on the environment by changing the produce they buy, how often they drive their car, recycling their waste, and curtailing aeroplane travel. Pet owners can also play a role in curtailing the damage that some pets do to the environment. Woods, McDonald, and Harris (2003) conducted a UK survey of the wildlife killed by domestic cats. With a sample of 618 households containing 696 cats, they found that the cats accounted for 14,370 animals: the most common victims were 20 species of mammals (69 percent of all prey), then 44 species of wild birds (24 percent), and three species of amphibian amphibians (4 percent); other prey included reptiles, fish, and invertebrates. Assuming a UK population of 9 million cats, the 5-month period spanned by the study of an estimated 92.1 million animals – 57.4 million mammals, 27.1 million birds, 4.8 million reptiles, and amphibians and 2.8 other animals – were killed by cats. Similar concerns about the predatory behaviour of cats are evident in other countries. Woinarski et al. (2017) reviewed 93 studies and estimated that an average of 272 million birds are killed annually in natural landscapes in Australia by feral cats, increasing to 377 million birds when those killed by feral cats in modified landscapes such as rubbish dumps and by pet cats are added to the total. This figure equates to an astonishing million birds a day.

In Australia, there is also concern about the impact of domestic and feral cats on the mammal population. Murphy et al. (2019) gathered data from 107 studies to estimate the number of mammals killed by cats across Australia. They concluded that feral cats account for 815 million mammals per year; the estimated Australian population of 3.88 million pet cats kill 180 million mammals every year. The devasting effects of cats on other wildlife is not, of course, confined to Australia. A Polish study by Krauze-Gryz, Gryz, and Żmihorski (2018) reported that cats kill over 500 million mammals and over 100 million birds. Linklater, Farnworth, van Heezik, Stafford, and MacDonald (2019) explain that in New Zealand cats are a particularly serious threat to birds and reptiles because much of the native fauna has evolved without mammalian predators. While the scale of the problem is evident, the solution is less so. Should biodiversity be protected by making it illegal to keep cats as pets or stipulating that they must be confined to the house? It would be a brave legislature that passed laws prohibiting feline pets and incurring the wrath of millions of cat owners.

In the same way that the demise of the dinosaurs opened the door for mammals, so the Anthropocene presents opportunities for other animals. We see animals such as feral cats colonising brownfield sites in cities while, as described by Giraud, Kershaw, Helliwell, and Hollin (2019), the common bedbug and the hookworm have found this to be a time of abundance.

Finally, where is psychology in all this change? Psychoanalysts have mused on what changes the Anthropocene may herald for therapy (LaMothe, 2020) and psychologists with a leaning towards critical psychology have reflected on the changes it must bring to humans and our relationship with other species (Adams, 2020). The concern of green psychology is with the need to change our relationship to the planet on which we live (Metzner, 1999). These are all positive steps and, looking on the bright side of life, mayhap that with genuine interdisciplinary efforts, it is not too late for animals and humans.

References

Adams, M. (2020). Anthropocene psychology: Being human in a more-then-human world. London: Routledge.

Crutzen, P. J., & Stoermer, E. F. (2000). The 'Anthropocene'. *Global Change Newsletter.* The International Geosphere-Biosphere Programme (IGBP): A Study of Global Change of the International Council for Science (ICSU), *41*, 17–18.

Foster, J. (2019). Let's not talk about the anthropocene. Analecta Hermeneutica, 10.

Giraud, E., Kershaw, E. H., Helliwell, R., & Hollin, G. (2019). Abundance in the anthropocene. The Sociological Review Monographs, 67, 357–373.

Krauze-Gryz, D., Gryz, J., & Żmihorski, M. (2018). Cats kill millions of vertebrates in Polish farmland annually. Global Ecology and Conservation, 17, e00516.

LaMothe, R. (2020). On being at home in the world: A psychoanalytic-political perspective on dwelling in the Anthropocene era. The Psychoanalytic Review, *107*, 123–151.

Lewis, S. L., & Maslin, M. A. (2015). Defining the anthropocene. Nature, 519(7542), 171–180.

Linklater, W. L., Farnworth, M. J., van Heezik, Y., Stafford, K. J., & MacDonald, E. A. (2019). Prioritizing cat-owner behaviors for a campaign to reduce wildlife depredation. Conservation Science and Practice, 1, e29.

Metzner, R. (1999). Green psychology: Transforming our relationship to the earth. Rochester, VT: Park Street Press.

Murphy, B. P., Woolley, L. A., Geyle, H. M., Legge, S. M., Palmer, R.,.... Woinarski, J. C. Z. (2019). Introduced cats (*Felis catus*) eating a continental fauna: The number of mammals killed in Australia. Biological Conservation, 237, 28–40.

Peters, A. (2020). Toward international animal rights. In A. Peters (Ed.), Studies in global animal law (pp. 109–120). Berlin: Springer Open.

Scanes, C. G. (2018). Human activity and habitat loss: Destruction, fragmentation, and degradation. In C. G. Scanes and S. R. Toukhsati (Eds.), Animals and human society (pp. 451–482). London: Academic Press.

Toukhsati, S. R. (2018). Animal extinctions. In C. G. Scanes and S. R. Toukhsati (Eds.), Animals and human society (pp. 499–518). London: Academic Press.

Woinarski, J. C. Z., Murphy, B. P., Legge, S. M., Garnett, S. T., Lawes, M. J., Comer, S.,... Woolley, L. A. (2017). How many birds are killed by cats in Australia? Biological Conservation, 214, 76–87.

Woods, M., McDonald, R. A., & Harris, S. (2003). Predation of wildlife by domestic cats *Felis catus* in Great Britain. Mammal Review, *33*, 174–188.

Epilogue

The joy of human–animal relationships is known to anyone with a family dog and we all have tales to tell of our dog's exploits. In the spirit of sharing, our present dog is a Cocker Spaniel called Toby who has been with us since puppyhood. When Toby was a few months old, he accompanied us on a short seaside break in Filey, Yorkshire. Taking a break from a walk along the front, we sat outdoors at a café watching the sun catch the waves, Toby sitting patiently at our feet. Coffee finished, we rose to leave and my wife took his lead and said, "Come on, Toby." As I followed, the women at the next table leant over to a little boy, perhaps aged 5 or 6, sitting next to her and I heard her say "That dog's got the same name as you!"

The small boy looked puzzled, paused, looked at his mam, and replied in a soft Yorkshire accent, "What, he's called Toby Gibson?"

Thus, Toby Gibson, shortened to TG, became Toby's sobriquet and our wider family will as a matter of course ask after TG.

Figure 10.1 Toby Gibson.
Source: Photograph by Clive Hollin.

Index

AAI (Animal Assisted Interventions) 147, 165
AAT (Animal Assisted Therapies) 147, 165–68
acute stress, in pets 77
addictions, therapy animals and 148
ADHD (attention deficit hyperactivity disorder), therapy animals and 152
adolescents; animal therapy and 151; criminal or potential criminal behaviour, animal therapy and 164; with mental health conditions, animal therapy and 153
aggression 52; in cats 29; in dogs 51–72; biting and 57–67; in male dogs vs. female dogs 51; physical punishment and 53; towards other dogs 54; towards people 55; off-leash dog walking and 37; punishing pets and 104; ways to reduce 91–98
ailurophobia 76
altruism, hunting and 184
American Animal Hospital Association (AAHA) 91
American Sign Language 12
American Society for Prevention of Cruelty of Animals (ASPCA) 33
animal-assisted detection 129–34; COVID-19 pandemic and 159; dog's olfactory system and 129
Animal Assisted Interventions (AAI) 147, 165
Animal Assisted Therapies (AAT) 147, 165–68
animal behaviour (ethology) 1
animal cognition 16
animal control, dog bites and 94
Animal Liberation 116
animal models 16
animal psychology 15
animal shelters 33
Animal Welfare Act 2006 88
animals; animal rights and 188; awareness of death and 42; enriching the lives of when in captivity 119; phobias of 76; used/harmed by humans 177–87; in warfare 134–39
antecedent, Skinner's three-term contingency and 10, 61

the Anthropocene 188
anthropomorphism 17
antidepressants 99
anxiety 73; animal therapy and 148, 167; in dogs 52, 73–80; ways to reduce 98–104
Anxiety Wrap 102
anxiety wraps 102
anxiolytics 99
apes, The Great Ape Project and 188
armadillo racing 178
ASPCA (American Society for Prevention of Cruelty of Animals) 33
assistance, provided by animals 126–39
assistance (service) dogs 127
Assistance Dogs International 126
astraphobia 77
attachment 11, 30; children, pets and 38; Harlow's research and 185; learned behaviour and 76; to animals in general vs. pets 39
attention deficit hyperactivity disorder (ADHD), therapy animals and 152
Australian Veterinary Association 88
autism, therapy animals and 149
autism spectrum disorder (ASD), therapy animals and 149

barking; anxiety and 73; bark-activated shock collars and 86; strategies to reduce 84
bears, in warfare 135
beasts of burden, warfare and 137
bed bugs; rising numbers of 189; search dogs and 132
behaviour; historical background of the study of 1; Skinner's three-term contingency and 10, 61
behaviour change; drug therapy and 99, 102; non-drug treatment and 103; principal components of success and 103; reward/punishment and 104; zoo visits/programs and 125
behaviour problems 51, 78; resolving 81–111; types of 51
Bekhterev, Vladimir Mikhailovich 5

birds 180; cognitive skills and 16; used in hunting 181
bites, by animals other than dogs 58
biting, by dogs 57–67; antecedents to 61, 95; children and 59; consequences of 64–67; defined 57; location/severity of bites 63; mental health disabilities and 60
blood sports 179
Blue Dog Programme 97
bonobos, The Great Ape Project and 188
breed-specific legislation (BSL) 94
breeding programmes, addressing behavioural problems and 81
breeding projects 188
British Veterinary Association (BVA) 33, 88
BSL (breed-specific legislation) 94
bullfighting 179
BVA (British Veterinary Association) 33, 88

cadaver dogs 130
camel wrestling 179
cancer, therapy animals and 154
Canine Behavioural Assessment and Research Questionnaire. *see* C-BARQ (Canine Behavioural Assessment and Research Questionnaire)
cardiovascular disease (CVD), dog ownership/ dog therapy and 155–58
castration, addressing behavioural problems and 81
Cat-Owner Relationship Scale (CORS) 27
cats 27–30; aggression and 29; allergies to 29; fencing containment systems and 88; historical perspective on 25; laboratory research and 7; owners and 28; phobia of 76; separation anxiety and 75; stress in 78; training 88; as warfare animals 138; wildlife killed by 189
C-BARQ (Canine Behavioural Assessment and Research Questionnaire) 56; aggression and 53; separation anxiety and 75
CBEs (Commercial Breeding Establishments) 32
Cecil, the lion 182
children; with autism, animal therapy and 150; childhood disorders, animal therapy and 151; courtroom dog assistance and 126; cruelty to animals and 178; dog bite prevention programmes for 97; dog bites and 59, 67; educational programmes and 125; with mental health conditions, animal therapy and 153; pets and 38
chimpanzees; laboratory research and 8, 185; The Great Ape Project and 188; Washoe and 12, 13, 16
chronic stress, in pets 77
circuses 116
clicker training 84
climate change 188
clomipramine 82

cockfighting 179
cognitive revolution 15
Commercial Breeding Establishments (CBEs) 32
companions. *see* pets
comparative psychology 15
confrontational/non-confrontational training methods 86
consciousness 16
consequence, Skinner's three-term contingency and 10, 61
conservation 184, 188; surveying wildlife and 131; zoos and 117, 122, 124, 125
contingency management, as a strategy to reduce aggression 91
continuous reinforcement, training and 83
CORS (Cat-Owner Relationship Scale) 27
courthousedogs.org 126
courtroom dogs 126
COVID-19 pandemic, detection dogs and 159
criminal offenders; animal therapy and 163; animal cruelty and 178
cruelty to animals 178, 180
CVD (cardiovascular disease), dog ownership/ dog therapy and 155–58
cyber hunting 181
cynophobia 76

The Dangerous Dogs Act 1991 92
DAP (dog-appeasing pheromones) 99, 100
deer 180
Defra (Department of Environment, Food and Rural Affairs) 88
Delta Society 147
Department of Environment, Food and Rural Affairs (Defra) 88
depression, effectiveness of animal therapy and 167
detection dogs 129–34; COVID-19 pandemic and 159; principal domains of 130
distal antecedents, dog bites and 61
distal consequences, dog bites and 65
dog-appeasing pheromone (DAP) 99, 100
Dog Bite Complication Index 63
Dog Bite Severity Ratings 64
dog bites, legislation/animal control and 94
dog paddocks 37
dog parks 37
dogs; aggression and 51–72, 96; anxiety in 73–80; assistance (service) dogs 127; courtroom assistance and 126; cruelty to 90, 178; dog walking and 36; dogfighting and 179; education assistance and 127; as food 177; high-volume breeding and 32; historical perspective on 25; laboratory research and; Ivan Petrovich Pavlov and 6, 184; Martin Seligman and 14, 184; mental health therapy and 148–54; phobia of 76; physical health animal therapy and 154–59; sense

of smell and 129; stress in 78; training, methods/effectiveness of 89; used in hunting 181; as warfare animals 138
dolphins 137
domestication of wild animals 25
dominionistic pet owners 30
drug detection dogs 134
drug therapy; for aggression problems 82; for anxiety problems 99; combined with behaviour management 99, 102; non-drug treatment and 102
DSM-V, psychological consequences of dog bites and 66
Dutch Socially Acceptable Behaviour Test 56

eagles 188
educational programs, zoos and 124
electric shock collars (e-collars) 86
electric/electronic devices 86; bans on/opposition to 87; cats and 88
emotional reactions, when visiting zoos 121
empathy, visiting zoos and 122
enriching the lives of captive animals 119
entertainment, animals as 115–26
environmental tourism 188
epilepsy, therapy animals and 155
ERWs (explosive remnants of war), search dogs and 133
ESVCE (European Society of Veterinary Clinical Ethology) 87
ethology (animal behaviour) 1
European Society of Veterinary Clinical Ethology (ESVCE) 87, 91
explosive remnants of war (ERWs), search dogs and 133

fence collars/fencing containment systems 86, 88
feral cats 189
field sports 179
financial consequences, dog bites and 67
fireworks, noise phobia and 77
fish; mental health therapy and 149; social information processed by 16
fishing 181
fluoxetine 82
food, animals as a source of 177
Fouts, Roger 12
foxes 180, 182
Freud, on animals 5

Gardner, Allen 12, 185
Gardner, Beatrix 12, 185
genotype, dog behaviour and 51
geological eras 188
Gibson, Toby 191
global change 188

gonadectomy, addressing behavioural problems and 81
gorillas, The Great Ape Project and 188
green psychology 189
greyhound racing 178
guide dogs 127, 147; attacked by dog 55; effects of ownership and 128
Guide Dogs for the Blind Association 127
Guide Dogs of America, The 127

Hahnemann, Samuel 101
Harlow, Harry 11, 185
harvesting 184
health benefits, dog walking and 37, 39, 156
hearing dogs 128
heart disease, dog ownership/dog therapy and 155–58
helicoptering a dog 90, 178
homelessness, pet ownership/animal therapy and 159–63
homeopathy therapy 101
hookworms, rising numbers of 189
horses; horse racing and 178; horse wrestling and 179; mental health therapy and 148–53, 166; as warfare animals 138
HSUS (Humane Society of the United States) 178
human-centric training, methods and effectiveness of 89
Humane Society of the United States (HSUS) 178
humanistic pet owners 30
hunting 179, 180; attitudes toward 182; methods of 181

imprinting 1
insects; as food 177; search dogs and 132
intermittent reinforcement, training and 83
internet hunting 181

Kennel Club 88, 90
Köhler, Wolfgang 8, 185
Kyoto Protocol 188

laboratory research; cats and 88; chimpanzees and 8, 12, 13, 16, 185; dogs and 6, 184; monkeys and 11; pigeons and 9; rats and 8, 184
language, teaching chimpanzees and 12
legislation, dangerous dogs and 94
lion hunting 182
Lorenz, Konrad Zacharias 1

Marine War Dog Memorial 139
mascots 134
medication. *see* drug therapy
mental health animal therapy 147, 161, 165;

effectiveness of 167; pet ownership and 159–63
mental health disabilities, dog bites and 60
metacognition 16
mobility assistance dogs 128
monkeys, laboratory research and 11

narcotics, search dogs and 134
negative punishment 10
negative reinforcement 10
neutering, addressing behavioural problems and 81
neutriceuticals 99
New Circus 116
noise phobia 76
Norwegian Scientific Committee for Food and Environment (Vitenskapskomiteen for mat og miljø (VKM)) 87
Nouveau Cirque 116

obsessive-compulsive behaviour; drug therapy and 99; stress and 78
off-leash dog walking 36, 37
olfactory system, in dogs 129
orangutans, The Great Ape Project and 188
ovariohysterectomy, addressing behavioural problems and 81
owner-directed aggression, dogs and 57
owners; of assistance (service) dogs, impacts and 128; behaviour by prior to dog bites 95; treatment integrity and 104

pain relief, animal therapy and 158, 165
Pavlov, Ivan Petrovich 5; associations, conditioning and 84; dog laboratory research and 6, 184
PBQ (Pet Bereavement Questionnaire) 41
People for the Ethical Treatment of Animals (PETA) 116
Pet Bereavement Questionnaire (PBQ) 41
pet owners; choosing a pet and 31; cultural differences and 28; statistics about 26; types of 30
pet stores 32
PETA (People for the Ethical Treatment of Animals) 116
pets 23–111; abandoned/euthanized 33; adopting 34; benefits of 31, 37, 40; death of/bereavement for 40; end-of-life decisions and 41; environmental damage caused by 189; lost/stolen 41; the owner-animal relationship and 30; problems with 42
phenotype, dog behaviour and 51
pheromone collars/diffusers 100
phobias 76
physical disabilities, assistance (service) dogs and 127
physical health animal therapy 154–59

pigeons; pigeon racing and 178; Skinner's research and 9; in warfare 136, 138
pollution 188
population control, hunting and 184
positive punishment 10
positive reinforcement 10
post-traumatic stress disorder. *see* PTSD (post-traumatic stress disorder)
prevention programmes 97
prohibition, as a strategy to reduce aggression 91
protectionist pet owners 31
proximal antecedents, dog bites and 61, 62
proximal consequences, dog bites and 65
psychological research, animals in 3–21
psychology; formative strands of 5; the Anthropocene and 189
PTSD (post-traumatic stress disorder) 66; dog bites and 66; effectiveness of animal therapy and 167; therapy animals and 149
punishing pets; vs. rewarding 89, 104; why it finds favour 90
puppies; anxiety in 75; socialization and 84; training 82
puppy farms/puppy mills 32

rabbits 180
rabies; in China 62; vaccination and 67
rage. *see* aggression
rats, laboratory research and 8, 184
recreational hunting 180
reinforcement, training and 83
remote-control hunting 181
rescue dogs, aggression and 61
RSPCA (Royal Society for the Prevention of Cruelty to Animals) 33; electric/electronic devices and 88; Pit Bulls and 93; reward-based training and 90

safaris 188
SAPS (Short Attachment to Pets Scale) 39
Scottish SPCA (Scottish Society for the Prevention of Cruelty to Animals) 39
search dogs 129; explosive devices and 133; inorganic substances and 133; narcotics and 134
The Seeing Eye 127
seizures, therapy animals and 155
Seligman, Martin 14, 184
separation anxiety 74–77; explanations for 75; noise phobia in dogs and 77
sertraline 82
service animals, vs. therapy animals 147
service (assistance) dogs 127
Shettleworth, S.J. 17
Short Attachment to Pets Scale (SAPS) 39
shuttle box 14
sight dogs 127

sign language 12
Skinner box 10, 184
Skinner, B. F. 9, 119, 184; dog attacks and 95; three-term contingency and 10, 61
Small, Willard S 8
social interactions; dog walking and 36; Harlow's research and 185
social well-being animal therapy 159
socialization, importance of early in life 11, 84
spectator sports 179
sports, animal used for 178–84
statistics, pet owners and 26
StopCircusSuffering.com 116
Storm Defender 102
stranger-directed aggression, dogs and 57
stress; animal shelters and 34; dog ownership benefits and 156; in pets 77
Sultan, chimpanzee research and 8
sustainability, hunting and 184
systematic desensitisation 103

termites, search dogs and 133
therapy animals 147–76; mental health therapy and 147–54; physical health therapy and 154–59; vs. service animals 147; social well-being therapy and 159
Theroux, Paul 1
Thorndike, Edward Lee 7
three-term contingency 10, 61
thunderstorms, noise phobia and 77, 102
Tinbergen, Nikolaas 1
Toby (TG) 191
training, methods and effectiveness of 82–91; aversion-based methods and 86; gadgets for 84; human-centric vs. dog-centric training and 89; reward-based methods and 83; reward vs. punishment 89; training cats and 88

trauma, effectiveness of animal therapy and 167
treatment integrity/treatment fidelity 104
trophy hunting 180, 183
TV shows, animals in 115

UFAW (Universities Federation for Animal Welfare) 88
unexploded ordnances, search dogs and 133
Universities Federation for Animal Welfare (UFAW) 88

vegans 178
vegetarians 178
vet visits, cats and 88
victims of dog bites, behaviour by 95
vivisection 88, 184
VKM (Norwegian Scientific Committee for Food and Environment (Vitenskapskomiteen for mat og miljø)) 87

Wagner, Vladimir Aleksandrovich 5
warfare animals 134–39; historical background of 134; memorials to the fallen 138
Washoe, chimpanzee research and 12, 13, 16
well-being animal therapy 159
whale watching 188
wildlife conservation, search dogs and 131
Wojtek, the bear 135
Wolpe, Joseph 103
wolves 188
Wundt, Wilhelm Maximilian 5

zookeepers' relationship with animals 119
zoos 117–26; educational opportunities and 124; visitor empathy and 122; visitors to, impact on animals 120